Sarah Julku

About the Author

JAMES L. NELSON has served as a seaman, rigger, boatswain, and officer on a number of sailing vessels. He is the author of *By Force of Arms*, *The Maddest Idea*, *The Continental Risque*, *Lords of the Ocean*, and *All the Brave Fellows*—the five books of his Revolution at Sea Saga, as well as *The Guardship*, *The Blackbirder*, and *The Pirate Round*—the books of the Brethren of the Coast trilogy. His most recent novel is *Glory in the Name: A Novel of the Confederate Navy*. He lives with his wife and children in Harpswell, Maine. His website is found at www.jameslnelson.com.

Also by James L. Nelson

The Revolution at Sea Saga
BY FORCE OF ARMS
THE MADDEST IDEA
THE CONTINENTAL RISQUE
LORDS OF THE OCEAN
ALL THE BRAVE FELLOWS

The Brethren of the Coast
THE GUARDSHIP
THE BLACKBIRDER
THE PIRATE ROUND

GLORY IN THE NAME: A NOVEL OF
THE CONFEDERATE NAVY

THIEVES OF MERCY: A NOVEL OF
THE CIVIL WAR AT SEA

James L. Nelson

Perennial

An Imprint of HarperCollins*Publishers*

THE WRECK OF THE IRON-CLAD "MONITOR."

REIGN

The Story of the First Battling Ironclads,
the Monitor and the Merrimack

of IRON

A hardcover edition of this book was published in 2004 by William Morrow, an imprint of HarperCollins Publishers.

HarperCollins books may be purchased for educational, business, or sales promotional use. For information please write: Special Markets Department, HarperCollins Publishers Inc., 10 East 53rd Street, New York, NY 10022.

First Perennial edition published 2005.

Designed by Jennifer Ann Daddio

The Library of Congress has catalogued the hardcover edition as follows:

Nelson, James L.
    Reign of iron : the story of the first battling ironclads, the Monitor and the Merrimack / James L. Nelson. — 1st ed.
    p.   cm.
    Includes bibliographical references and index.
    ISBN: 0-06-052403-0 (acid-free paper)
    1. Monitor (Ironclad). 2. Virginia (Ironclad). 3. Merrimack (Frigate). 4. United States—History—Civil War, 1861–1865—Naval operations. 5. Hampton Roads, Battle of, Va., 1862. I. Title.
E595.M7N45 2004
973.7'52—dc22

2003064938

ISBN 0-06-052404-9 (pbk.)

05 06 07 08 09 ❖/RRD 10 9 8 7 6 5 4 3 2 1

TO NATHANIEL JAMES NELSON,

my first son, my beloved boy

Now comes the reign of iron—and cased sloops
are to take the place of wooden ships.

—REAR ADMIRAL JOHN DAHLGREN,
COMMENTING ON THE FIGHT BETWEEN
MONITOR AND *VIRGINIA*

Contents

CONTENTS

Acknowledgments

The writer of history, I have discovered, must depend greatly upon the kindness of strangers, and I am indebted to many for their help. The Mariner's Museum in Newport News, Virginia, is the epicenter of all things *Monitor*, and much material regarding *Virginia* as well, and the staff could not have been more helpful. My sincere thanks to the library staff, Josh Graml, Marc Nucup, Lynnette Accors, Lester Weber, Gregg Cina, and especially Cathy Williamson. Thanks also to the turret's conservator, T. R. Randoph, for showing me around that extraordinary artifact.

The National Oceanic and Atmospheric Administration's Jeff Johnson kindly answered my nitpicky questions regarding *Monitor*, and I am indebted to him for that.

At the United States Naval Academy's Nimitz Library, a thank you to Jennifer Bryan, Head of Special Collections and Archive, and Mary R. Catalfamu, Librarian, Special Collections. Claire McKillip at the Allis-Bushnell House in Madison, Connecticut, was a great help. Thanks to John Rhodehamel at the Huntington Library, Marjorie McNinch at the Hagley Museum & Library, in Wilmington, Delaware, and the folks at the University of North Carolina, Chapel Hill, and the Virginia Historical Society in Richmond.

I am grateful to the staff at the Bowdoin College Library, that wonderful institution that allows me to live in a small coastal town and still have access to a first-rate library. Even more helpful on that count were the research librarians at the Curtis Memorial Library in Brunswick, Maine, and I thank them for their tireless efforts in securing material through interlibrary loan. They are Janet Fullerton, department head; Linda Oliver, Sally Jeanne Kappler, Michael Arnold, and Laura Bean. Thanks also to Dave Clark at the Folger Library at the University of Maine, Orono, and Nathan Lipfert, librarian at the Maine Maritime Museum.

Thank you to Joan Druett, for all her help and advice in this strange new world of historiography, and to Van Reid for his ongoing friendship and support.

Bill Whorton kindly gave me the benefit of his vast knowledge of the Confederate Navy. I am indebted, as I have been for all of my writing that involves steam engines, to Ed Donohoe, who may think like an engineer but who spins a yarn like a sailor. Thanks as well to Robin Walbridge, captain of the sailing ship *Bounty*, for all his support and for sailing me past the places where *Merrimack* became *Virginia*.

On the publishing front, I am grateful as always for the support of my editor, Hugh Van Dusen, and to David Semanki for all his tireless help. Thank you to Dee Dee DeBartlo for all her work. And most of all (as usual) thanks to Nat Sobel, who was the prime mover in making this and so much else happen.

Introduction

If you go to Newport News, Virginia, today you can do something quite unique—drive over the site of a naval battle. Where the first battle of ironclads took place, the five-mile-long Monitor-Merrimac Memorial Bridge Tunnel now connects Newport News with Portsmouth to the south. About midway across the mouth of the James River, the bridge plunges down into a tunnel, emerging at just about the exact point where the U.S.S. *Congress* met her blazing death at hands of Confederate ironclad sailors.

Driving across that amazing bit of engineering, long and low, a straight road thirty feet above the water, illuminated at night by brilliant streetlamps, one wonders what the men of the *Monitor* or *Merrimack* would think, could they see it. Would they be dumbfounded at the engineering that went into a four-lane highway that runs under the water? Would they be shocked by the technology that allows people to drive combustion-engine–powered vehicles at sixty miles an hour over the bridge, under a blaze of artificial light?

Probably not.

Certainly the technological advances might amaze them, but the technology itself would not be beyond their comprehension. The men who sailed the *Monitor* and the *Merrimack* lived in a world vastly different from that of a generation before. The industrial revolution was no longer in its sputtering early days, but was firmly entrenched. Iron and steam were there to stay.

The men who sailed the ironclads were manning the most technologically advanced ships of their day. They understood the direction where technology was going. They embraced the new and the innovative. They were not mystified or afraid of the future—they were making it.

Monitor and *Virginia* were not the first ironclads. They were not even the first American ironclads, on either the Union or Confederate side. They were the first ironclads to fight one another. They were the first ships to demonstrate what an ironclad could do, and in doing so, they revealed to a skeptical world how every other ship seemed paltry in comparison.

Each ship taught the world a unique and valuable lesson. *Virginia*, in her first day of battle, proved conclusively that wooden vessels were obsolete in the presence of an ironclad man-of-war. *Monitor* proved that an ironclad could fight an ironclad, and if she did not actually beat *Virginia*, she showed that with the right combination of armor and guns it could be done. More importantly, she showed that the new technology required an entirely new way of thinking about ship design. The old man-of-war paradigm, with high sides and guns arranged to fire in broadside, such as *Virginia* still sported, would soon be as antiquated as the wooden walls.

Merrimack, Merrimac, or *Virginia*?

The ship that was refitted as the Confederate States Ship *Virginia* was the old United States Ship *Merrimack*. Even before the conversion, indeed even before the original ship was built, there was confusion as to the spelling of her name. On the sail plan drawn up for the ship's construction, her name is written as *Merrimack*, while the drawing of her lines and the deck arrangement show the name as *Merrimac*.

Merrimack was the first of six screw frigates begun in 1854. She, like the others in her class, the *Wabash, Roanoke, Niagara, Minnesota,* and *Colorado*, were named after American rivers. In Massachusetts, the Merrimack River runs through the better-known Merrimac Valley, and therein lies the confusion. Though the USS *Merrimack*'s name was spelled as often without the "k" as with, it is, properly, *Merrimack*.

When the steam frigate fell into Confederate hands, she was still the *Merrimack*. Virtually all of the officers in the Confederate navy had been officers in the Union navy, and they knew the old ship well by her original name. In a few instances she is even referred to as the CSS (Confederate States Ship) *Merrimack*.

Sometime in late 1861 it was decided that the old USS *Merrimack* would be renamed CSS *Virginia*. On February 17, 1862, she was commissioned and the name became official. From that point on, she was virtually never referred to as *Merrimack* in any Confederate documents, official or private. One of many examples of this switch is in the ship's payroll records. The account book for January 1862 lists "appropriations for the '*Steamer* Merrimac'." The accounts for February are for the *Virginia*.

No one on the Union side ever called her the *Virginia*. She was always the *Merrimack*. Under the heading of "the winners write the history," in the postwar years she continued to be referred to as the *Merrimack*, to the point where a number of Confederates, writing about the war in later years, also referred to her as the *Merrimack*, no doubt to avoid confusion. This has led to the mistaken impression that neither side ever called her the *Virginia*, which is not the case. After February 1862, she was the *Virginia* in the South.

William Cline, a crew member onboard the *Virginia*, resisted the Northern revisionist renaming. In an article written forty-two years after the battle he wrote, "History, in all cases that I have heard of, refers to the ship as the *Merrimac*, but I want to say right here that there never was a vessel in the Confederate States navy called by that name."

It is easy to see how the ship was referred to in the written record, harder to know what she was called in casual conversation, and in that it seems she may still have been called the *Merrimack* by her new owners. E. A. Jack, one of *Virginia*'s engineers, referred to "the 'Merrimac' now called the 'Virginia', but except in official communications, ever called by the former name." John Wise, recollecting the ironclad that, as a boy, he had seen built in the nearby Gosport Naval Shipyard, remembered always calling her the *Merrimack*. "Such we called her, for we had never become accustomed to the new name, *Virginia*."

The reality, no doubt, was that some called her *Merrimack* and some called her *Virginia*, the two as interchangeable as the First Battle of Manassas and the First Battle of Bull Run. In this work, I have tried to remain true to the contemporary names for the ship; thus she is *Merrimack* up until her rechristening, and *Virginia* thereafter.

Chapter 1

Sink Before Surrender

Saturday, March 8, 1862, was a beautiful day in Virginia. A gale had blown itself out the night before, and behind it came clear, warm weather, a high-pressure system on the tail of the storm. A day more like May than March, many people felt.

In Norfolk and Portsmouth, towns that faced one another across the Elizabeth River, excitement was spreading like fire, just as it had the year before, in the days leading up to secession and the burning of the shipyard. The Confederate States Ship *Virginia*, an ironclad built on the burned-out hull of the old USS *Merrimack*, was getting under way. There had been no announcement. In the interest of military security, the Gosport Naval Shipyard had been closed to visitors for months. Not even *Virginia*'s crew knew where they were bound.

But there was no concealing her movements. *Virginia* was a monstrous vessel, 275 feet long. She was 38½ feet on the beam, and though the crowds watching from the shore could not see this, she was burdened by a ponderous 22 feet of draft.

With black smoke rolling out of her tall stack she edged away from the dock, heading into the stream. Word spread fast, and people rushed to the riverbank to see her go. They had been waiting eight months for this moment.

Most of *Virginia* was underwater, not only her massive hull, but also her afterdeck, the last 50 feet or so of the ship, which was 6 inches below the surface. All that the citizens watching could see was a wedge-shaped false bow, barely breaking the surface, and her ironclad shield, like a barn roof floating on the river, 8 feet high. The lengths of plate iron running vertically along the shield gleamed black with the coat of tallow smeared on them to help enemy shot bounce off. On the forward flagstaff flew the red pennant of an admiral. On the ensign staff was the Confederate national flag, the "Stars and Bars."

The roof of the casemate, the "shield deck," was mainly an iron grating to let air and light into the gun deck below. But still the gun deck was "badly ventilated, very uncomfortable," and so gloomy that lanterns were needed the full length of the deck, even on a fine, sunny day such as the 8th.

For that reason most of the *Virginia*'s crew were crowded on the shield deck, about 16 feet wide and 120 feet long. In keeping with traditions of the sailing navy—men before the mast and officers aft—the crew stood in front of the smokestack, the officers aft of it, though the helm and pilothouse were at the forward end of the casemate.

Foremost of the officers was Franklin Buchanan, appointed admiral in command of the James River squadron just a few weeks before. Sixty-one years old, balding with a tussle of white hair ringing his head, Buchanan was a hard-driving disciplinarian, navy to the marrow, the "beau ideal of a naval officer of the old school, with his tall form, harsh features and clear piercing eyes." He was a man with a great deal on his mind.

Virginia had never been under way before. She was powered by the *Merrimack*'s old engines, engines that had been condemned by the U.S. Navy. Her engineer, H. Ashton Ramsay, had served aboard the ship while she was still the USS *Merrimack*, and he reported, "From my past and present experience with the engines of this vessel, I am of the opinion that they can not be relied upon. During a cruise of two years . . . they were continually breaking down, at times when least expected."

Buchanan had quizzed Ramsay about the engines before getting under way. He asked about their reliability. He asked how they would endure the shock of *Virginia* ramming another vessel. He asked if they should first make a trial trip.

Ramsay answered as best as he could. "She will have to travel some ten miles down the river before we get to the [Hampton] Roads. If any trouble develops, I'll report it. That will be sufficient trial trip."

But Buchanan had more than engines to worry about. The crew were new to the ship. Construction had been ongoing until the very end—that very morning he had ordered workmen off the ship so she could get under way—and the men had had no chance to drill onboard. They had never fired the guns. "The officers and crew were strangers to the ship and to each other," one of *Virginia*'s lieutenants wrote.

Many of the crew were strangers to ships of any description. The South had a chronic dearth of sailors, and *Virginia*'s men had been hustled from the army or recruited from among the yard workers or from local militia units. Scattered among them were a few veteran sailors, some survivors of the desperate battle for Albemarle Sound. "They proved to be as gallant and trusty a body of men as anyone could wish to command," recalled Midshipman Virginius Newton, "but what a contrast they made to a crew of trained jack tars!"

Virginia was a "novelty in naval construction," her properties unknown, and she was still incomplete. There had been no time to fit the protective shutters over the gunports. The ship was riding too high in the water. The lower edge of her casemate, which was supposed to be two feet underwater, was only a few inches under, leaving her lightly armored waterline vulnerable.

The enemy had at least five major warships on station, protected by heavy shore batteries at Newport News and the guns of Fortress Monroe and Fort Wool.

Any commanding officer would have been excused for insisting on a sea trial, a shakedown, a practice run, before steaming into battle. Most of the men onboard *Virginia* assumed that was what they were doing. Only a few knew the truth.

Buchanan was not going to wait. He was an aggressive fighting man with a bitter resentment of the U.S. Navy and he knew that the strategic situation would not allow for delay. Union General George McClellan was planning an attack on Richmond that he would launch from

Fortress Monroe, transporting his troops by water from Washington. If *Virginia* ruled Hampton Roads, she could ruin his entire plan.

And, more ominously, there were indications that a Union ironclad would soon be ready for sea.

So *Virginia* steamed away, bound for battle on her maiden voyage. The wharves along the Elizabeth River, the banks, the rooftops were crowded with people. They waved their hats and handkerchiefs, and the ironclad's crew doffed their hats in reply.

Some of *Virginia*'s men recalled the people onshore cheering loudly as the ironclad steamed past, shouting "Godspeed" as they headed down-river. Most remembered a silent, somber crowd who watched and waved but did not cheer. The Confederacy had suffered many setbacks in the past months after the jubilation of its victory at Manassas the summer before. A great deal of hope rested on the ship, but many believed she and her crew would not survive the day. The churches were jammed with people praying for the safety of the men and the success of their mission.

As the *Virginia* passed Norfolk, the hands were piped to dinner. In the engine room, Ramsay nervously watched as the two horizontal, back-acting engines turned under the pressure of the steam from the four huge Martin-type boilers. Everything seemed to be in order. He went back up to the shield deck.

"How fast is she going, do you think?" he asked one of the pilots.

The pilot looked at the shore and estimated the speed. "Eight or nine knots an hour," he replied, a very optimistic guess. It took the ironclad an hour and a half to steam approximately seven miles to the mouth of the Elizabeth River, putting her speed at closer to five knots, which was still respectable, given that she was underpowered, unwieldy, and stemming a flood tide. Ramsay and William H. Parker, the captain of the *Beaufort's,* one of *Virginia*'s escorts would later agree that she averaged around seven knots. That was as good as she did under power as a sailing frigate.

The crew were not as impressed with her turn of speed. "If this is all the speed we can make," they whispered to one another, "we better get out and walk."

Dinner was a brief affair, and terminated by the drums beating to quarters. The men scrambled to clear the ship for action, the first time it had ever been done. It was no doubt a confused and disorganized exercise, compared to a well-drilled company of bluewater sailors on board a man-of-war in long commission.

The galley fires were extinguished, lashings on the guns cast off, rammers and swabs arranged for easy handling. The powder magazine was opened and cartridges were passed up to the gun crews and rammed down the gaping barrels. On top of the charges went 9-inch shells for the Dahlgren smoothbores in the broadside, 6.4-inch shells for the Brooke rifles amidships, and 7-inch shells for the Brooke rifles in the bow and stern.

As the *Virginia* steamed past the Confederate batteries on Craney Island, the troops there lined the parapets and cheered wildly. By then there was a veritable parade of boats on the river, and "everything that would float, from the Army tug-boat to the oysterman's skiff" was loaded with spectators heading down to Craney Island to watch the action.

Buchanan ordered all hands topside. The shield deck was crowded with the 320 or so members of the ship's company, officers to starboard, enlisted men to port, waiting to hear what "Old Buck" would say.

Among those who recalled Buchanan's words, there seems to be little consensus about what he said. Third Assistant Engineer E. Alex Jack recalled Buchanan assuring those who might have had doubts about his loyalty to the Confederacy that after that day there would be "no cause for any such unjust suspicions," hardly words to inspire men to fight.

Another version had Buchanan saying, "The Confederacy expects every man to do his duty," but one hopes for a little more originality than that.

H. Ashton Ramsay's version was probably closer to the truth: "Sailors, in a few moments you will have the long expected opportunity to show your devotion to our cause. Remember that you are about to strike for your country, for your homes, for your wives and your children."

Ramsay, worrying over his engines, had nearly missed dinner. The caterer sent word to him that he had better eat soon or he would miss his chance. He climbed down to the gun deck, struck by the looks of the men waiting at the guns, "pale and determined, standing straight and stiff, showing their nerves were wrought to a high degree of tension." They stood with rammers and sponges in hand, waiting for the battle to start. Ramsay had been so involved in the engine room that he had not until that moment considered the bloody fight in which they were about to engage.

Down in the wardroom he found several officers "daintily partaking of cold tongue and biscuit." At the far end of the room, Assistant Surgeon A. S. Garnett was laying out tourniquets, forceps, bone saws, all the tools of his work. Ramsay's appetite disappeared. He settled for a taste of tongue and a cup of coffee.

Virginia was not going into battle alone. She was, in fact, the flagship of a squadron, the James River squadron. Three of the squadron's six vessels were up the James River, waiting for *Virginia* to make her move. In company with the ironclad were the *Raleigh* and the *Beaufort*, both converted tugboats mounting 32-pounder guns on their bows. They were instructed to use their "best exertions to injure or destroy the enemy."

For the time being, they were more useful as tugs. As *Virginia* passed Craney Island, the water became so shallow that she could not steer well. *Beaufort* passed a hawser and took her in tow until they reached the deeper water abreast Sewells Point.

Hampton Roads is a wide expanse of water and from the shore looks perfectly navigable in any direction. But that is deceptive. Most of the Roads are mud banks and shallow places. Like a wagon stuck in the ruts of a road, the deep-draft *Virginia* was limited in where she could go. To get at the enemy she would have to travel a roundabout course that would cover twice the distance of a direct line. There was no chance for the surprise Buchanan had originally hoped for.

As *Virginia* passed Sewell's Point, the men could see the enemy on either beam. To the east was Fortress Monroe and the men-of-war under its guns, *Roanoke*, commanded by Captain John Marston, senior captain present, with forty guns; *Minnesota*, with forty-eight guns; and *Brandywine* and *St. Lawrence*, both fifty-gun sailing frigates. To the west, blockading the entrance to the James River, were the sailing ships *Congress*, fifty guns, and *Cumberland*, thirty. Buchanan had already made up his mind on his initial attack. He turned the *Virginia* west.

Bent on to halyard on the forward flagstaff, ready to hoist, was the signal flag Number One. Buchanan had prearranged it with the other captains of the squadron. Number One hoisted below his pennant meant "Sink before you surrender."

"That Thing Is A-comin' . . ."

The Union ships *Congress* and *Cumberland* had been waiting a long time, at least since November of the previous year, for the day the Confederate ironclad would come out. They had spent the winter at anchor off Newport News, at the mouth of the James River, their time occupied by shipboard routine, drills, boredom, and misery.

It was no secret that the Confederates were building an ironclad on the raised hulk of the *Merrimack*. Rumors of her imminent appearance had surfaced so often that it was a standing joke onboard the ships. But joke or not, a high level of vigilance was always maintained. Regular sea watches were kept, with one watch always standing by or sleeping at the guns, and not the more relaxed system of anchor watch, which involved fewer and less vigilant men. Covered in blankets and pea jackets, the gun crews would often wake to find themselves under a layer of snow.

No hammocks were allowed on the gun deck, so that time would not be wasted getting them down in case of a fight. The guns were always loaded, and at sunset they were cast off and primed. No lights were shown above the waterline and fires were not allowed onboard, making the cold winter even more miserable. But the ironclad did not appear.

Still, by early March, the U.S. Navy had concluded that the ships in Hampton Roads were too vulnerable to attack by the ironclad and decided to move a number of them to Washington. Navy Secretary Gideon Welles telegraphed instructions to John Marston, senior captain:

> Send the *St. Lawrence, Congress,* and *Cumberland* into the Potomac River. Let the disposition of the remainder of the vessels at Hampton Roads be made according to your best judgment after consultation with General Wool. Use steam to tow them up. I will also try and send a couple of steamers from Baltimore to assist. Let there be no delay.

Gustavus Fox, assistant secretary of the navy, began organizing steam tugs to tow the sailing ships up the river, then headed for Hampton Roads to make the arrangements himself. Gideon Welles telegraphed

Marston again, instructing him not to move any ships until Fox arrived. The day he sent those orders was March 8.

The beautiful early spring weather was a godsend to the men on blockade duty, particularly after the wicked storm that had just blown through. The ships were riding at an easy anchor, spring lines rigged to allow them to swing their broadsides in whatever direction they chose. The sails were loosened off to dry.

The morning watch on Saturdays was wash day, and the crews of *Congress* and *Cumberland* had taken advantage of the fine weather to perform that chore. Laundry lines were run up between the mainmast and the mizzen and clothes hung out to dry, white clothes to starboard, blue to port, per navy tradition. The ships' boats were in the water and tied to the lower studdingsail booms.

It was as close to a lazy, carefree day as the men would get. There was a relaxed attitude onboard, particularly onboard *Congress*, which, in just a few days' time, was to be sent north, relieved of the tedium of blockade duty at anchor. It was the navy's intention to replace both ships soon. In the age of steam, sailing vessels were of little use on blockade.

The surgeon onboard the *Congress*, Edward Shippen, took advantage of the fine weather to stroll along the poop deck, watching the gulls fighting for dinner scraps and exchanging words with Erie Kemp, the officer on deck. *Congress* had been on station in Brazil before being called back to the United States at the outbreak of the war. For Shippen, the cold winter they had endured at the mouth of the James was an unpleasant contrast to the previous winter "in the land of bananas and cocoanuts."

Few of the officers from that time were left, and almost none of the crew. On January 13, three hundred of *Congress*'s men had been paid off and discharged. To make up for the loss, troops from the 99th New York Infantry, the Union Coast Guard, had been stationed on the ship. For a month and a half they had drilled at the great guns and waited for the alleged ironclad to come out.

It was around twenty minutes before one o'clock in the afternoon when Shippen noticed the quartermaster and Kemp staring intently east toward the Norfolk Channel. The quartermaster turned to Shippen and handed him a telescope. "I wish you would take the glass and have a look over there, sir," he said. "I believe that thing is a-comin' down at last."

Shippen looked through the glass. "There was a huge black roof, with a smoke-stack emerging from it," he recalled, "creeping down towards Sewell's Point."

Sewell's Point hid the *Virginia* from the sight of the ships at Fortress Monroe, but not from Newport News. The officers onboard *Congress* gathered aft to watch the ironclad's approach.

It was not long before *Virginia* reached Hampton Roads and turned west toward the James River channel. *Congress* and *Cumberland* wasted no time in reacting. Drums beat to quarters. The laundry came down from aloft on a run, the boats were let loose from the booms, and the sails were furled. The magazines were opened and powder monkeys ran cartridges up to their guns. Powder and solid shot were rammed home. And then there was nothing more to do but wait.

To make the wait even more maddening, the *Virginia* was soon lost to sight from the two ships, disappearing behind a bluff at Newport News Point. For more than an hour and a half, the ships' companies waited, the men "standing at their guns for the last time; cool, grim, silent and determined Yankee seamen, the embodiment of power, grit and confidence."

They might well be confident. They were among the best-trained, most experienced man-of-war's men in the world, the equal of any navy of any nation. Their ship was armed with a powerful battery, which they could serve with speed and accuracy.

But in the history of naval combat, no wooden ship of war had ever fought a vessel the likes of *Virginia*. The Yankee sailors were confident because they had no idea what they were up against.

Iron Versus Wood

Onboard the *Virginia*, they could see the effect their appearance was having on the blockading fleet. They saw the wash onboard *Congress* and *Cumberland* come sailing down, the boats cast off, sails furled, all the evidence of ships clearing for action. The smaller vessels going about their business on the Roads "scattered like chickens at the approach of a hovering hawk." Smoke began to pour from the stacks of

Minnesota and *Roanoke* as their engineers flung combustibles on the fires to start the water in the boilers boiling and "get a head up steam."

Beaufort, coming astern of *Virginia*, fired first from long range, but her shot fell short. For this action—opening fire before the flagship—and others, Buchanan would report that William Harwar Parker was "unfit for command."

Buchanan held his fire until he was less than a mile away. It was 2:10 P.M. The tide had turned and the *Congress*, the nearest of the two ships, had swung stern to the approaching enemy. Only her after guns would bear, so the gunners opened fire with those. The solid shot struck the front of *Virginia's* casemate and bounced away like "water from a duck's back."

With those opening shots, all of the Union firepower in the theater opened up on *Virginia*, a massive barrage. The *Congress*, the *Cumberland*, the shore batteries, the small Union gunboats all poured fire into the ironclad, creating "a veritable storm of shells which must have sunk any ship then afloat." Over one hundred heavy guns were concentrating their fire on *Virginia*, but the Confederate ironclad steamed through the hail of iron as if it were a summer rainstorm.

Congress was the closer ship, but Buchanan had his eyes on *Cumberland*. In terms of the number of guns, *Congress* was the better armed, but she had mostly old 32-pounders in her broadside, while *Cumberland* had 9-inch shell guns and 11-inch pivot guns at the bow and stern, these latter of particular concern to the Confederates.

Virginia's first shot came from the 7-inch Brooke rifle in the bow. Lieutenant Charles Simms, hunched over the gun, trained it on *Cumberland*, stepped aside, and jerked the lanyard. The shell tore through the *Cumberland's* bulwark and exploded on the after pivot gun, killing and wounding ten men on the gun crew.

Buchanan called Ramsay back to the pilothouse and reiterated his intention of ramming the *Cumberland*. He ordered Ramsay to back the engines as soon as he felt the concussion, not to wait for orders. Ramsay acknowledged the instructions and returned to the twilight world of the engine room.

On the gun deck, Lieutenant John R. Eggleston was peering out of the midships gunport. He was in charge of the two "hot shot" guns, the guns designated to fire solid shot, heated red-hot in the furnaces below. With the limited view through the gunport he could see only water and the distant shore. Then, suddenly, Congress was there, only 100 yards away, filling the port.

Congress fired, her entire broadside, "the flash of thirty-five guns," nearly 1,000 pounds of iron slamming into Virginia at point-blank range. To Eggleston's relief, none of the shot came in through the gunports.

The noise inside the iron casemate was terrible. But when the men expressed concern, Lieutenant Charles Simms shut them up, telling them, "Be quiet, men, I have received as heavy a fire in open air." And that was the difference. Every shot that Congress fired had bounced off the shield and done no damage at all. Ironclad technology had passed its first test.

"A Slaughter-Pen"

Onboard Congress, they watched with dismay as their powerful broadside pelted off Virginia's thick hide and "rattled from the sloping armor like hail upon a roof." Then it was their turn to receive fire.

Virginia let off a full broadside as she passed, three 9-inch Dahlgrens and a 6.4-inch Brooke rifle, all loaded with explosive shell. The effect was murderous, turning the gun deck into a "slaughter-pen." An 8-inch gun was dismounted, its crew wiped out, killed or wounded. Shippen was stunned by the carnage around him, "lopped off arms and legs and bleeding, blackened bodies scattered by the shells, while blood and brains actually dripped from the beams."

He saw one man skewered by a wood splinter as thick as his wrist, but splinters, the old nemesis of sailors aboard wooden men-of-war, now paled in comparison to the butchery doled out by exploding ordnance. The shells left few wounded men, instead killing them outright. Ernie Kemp lay by the wheel with both legs shot off, calling with his last breath for the men to stand by the ship.

Virginia delivered her broadside and steamed on past, making for her intended target. On Cumberland they stood silently at their guns, watching with horror the beating that Congress received, but they did not lose heart. The destruction "caused us neither surprise nor shaken confidence in our own powers," *Cumberland*'s Lieutenant Thomas Selfridge recalled, "since the *Congress*'s armament could fire nothing to compare with the solid shot of 80 pounds which we could deliver." They still had no sense for the near-invincibility of *Virginia*'s wood and iron casemate.

But as *Virginia* approached, *Cumberland* could not bring any guns to bear. The ebbing tide had swung them "athwart the stream," broadside to the current, and *Virginia* was approaching from the bow. The only gun that would bear was the forward pivot, but if they fired that they would have blown their own head rigging away.

Cumberland's captain, Commander William Radford, was on board USS *Roanoke*, 7 miles away, sitting on a court of enquiry, leaving first officer Lieutenant George U. Morris in command. Morris ordered hands to take up on the spring line to swing the ship broadside to her attacker. On the spar deck, the men heaved at the capstan. But the tide had swung the ship in such a way that the spring line was running parallel to the keel, and hauling it would not make the vessel turn. The big guns they were counting on to destroy the Confederate ironclad would not even bear.

Virginia lay about 300 yards off *Cumberland*'s starboard bow, blasting the Union ship with her broadsides and bow gun. The first shot tore through *Cumberland*'s hammock netting, killing and wounding nine marines. The screams and groans of the wounded were something new to the men onboard, most of whom had never before been in combat.

Lieutenant Simms aimed *Virginia*'s bow gun. Having fired the gun that wiped out *Cumberland*'s after pivot gun's crew, he now did the same to the men on the forward pivot. The blast from the "murderous 7-inch rifle burst among the crew as they were running the gun out . . . literally destroying the whole crew except for the powder boy, and disabling the gun for the remainder of the action." The gun captain, training the gun with a handspike, had both arms taken off at the shoulders.

Virginia poured the shot into *Cumberland* and *Cumberland* was all but helpless to hit back. "[W]e could reply only by extreme train with the few guns," Selfridge wrote. "It was a situation to shake the highest courage and the best discipline, but our crew never faltered."

The *Cumberland*'s gun crews stood at their guns, firing as they could, while *Virginia* tore ship and men apart. The dead were flung over to the port side of the deck, the wounded carried down below, piling up too fast for the surgeon to keep up. As men were killed at the guns, others stepped up to take their place, as they had been trained to do.

Finally *Virginia*'s bearing shifted and more of *Cumberland*'s guns would bear. The first gun division had lost all of its first and second gun captains. Lieutenant Selfridge, in command of the first division, went from gun to gun, a box of cannon primers in his pocket, firing each gun as it was loaded and run out.

The devastation onboard *Cumberland* was horrible, shells and splinters cutting the men down where they stood, the deck running with blood. Their vaunted 80-pound shot was useless against the ironclad. But it was not Buchanan's intention to waste precious ammunition destroying the Union ship.

Virginia turned bow-on toward *Cumberland*, her engines making turns for full speed ahead. Richard Curtis, part of Lieutenant Simms' division on the bow gun, stared out the gun port at the frigate looming in front of them. *Cumberland*'s starboard side was "lined with officers and men with rifles and boarding pikes, all ready to repel us, thinking we intended to board her . . ."

Curtis had only a glimpse of the *Cumberland*'s men before *Virginia*'s iron ram slammed into the frigate's hull, right on the starboard bow, just under the starboard forechains, with enough force to open up a hole "wide enough to drive in a horse and cart."

Heavy wooden spars had been arranged around *Cumberland*'s bow to ward off floating explosives, but *Virginia* smashed those like matchwood. The "sound of crashing timbers was distinctly heard above the din of battle." From thirty feet away, Simms fired his bow gun into the packed men at *Cumberland*'s rail.

On *Virginia*'s gun deck, the collision was hardly felt. "The shock to us on striking was slight," Catesby Jones recalled. Others said the blow was "hardly perceptible," or "as if the ship had struck ground." The flag lieutenant, Robert Minor, rushed along the gun deck waving his hat and shouting, "We've sunk the *Cumberland*!" In the dim-lit casemate, choked with smoke, their view limited to the small gunports, most men aboard did not know what was going on.

Buchanan did not forget the engine room signals. In the last instant he feared they would hit *Cumberland* too hard, perhaps inextricably wedge themselves in the Union ship's side. As the *Virginia* raced down on *Cumberland*, seconds before impact, he rang up two gongs, the signal to stop engines, followed immediately by three gongs, full astern. And then the ship struck.

The impact was felt more strongly in the engine room. E. A. Jack felt a tremor run through the ship, and "was nearly thrown from the coal bucket upon which I was sitting." Ramsay recalled "an ominous pause, then a crash, shaking us all off our feet."

Cumberland rolled over hard with the impact. "I could hardly believe my senses when I saw the masts of the *Cumberland* begin to sway wildly," one witness recalled. As she rolled, she came down on *Virginia*'s iron ram, embedded in her side. In the engine room, Ramsay felt the bow being pushed down as the engines pounded in reverse, struggling to free the ironclad from the dying ship.

On *Cumberland*'s deck, Lieutenant Selfridge understood what was happening. "As the *Cumberland* commenced to sink, the *Merrimac* was also carried down until her forward deck was underwater." *Cumberland* was going down, and if *Virginia* did not free herself from her victim's side, she would be dragged down with her.

Chapter 2

Sea Trials of
the *Monitor*

It is very important that you should say exactly the
day the Monitor *can be at Hampton Roads.*

—GUSTAVUS VASA FOX TO JOHN ERICSSON

Forty hours before CSS *Virginia* began her rampage through the Union fleet, the only ship in the Western Hemisphere that might have stopped her was on the verge of sinking into the North Atlantic.

On March 3, the ironclad USS *Monitor* got under way from the Brooklyn Navy Yard to test a new arrangement of her steering gear, the last serious obstacle before she could steam south for Hampton Roads. The steering proved satisfactory, but for two days following the test the ship was kept dockside by foul weather. It was the same storm, sweeping the Atlantic coast, that had kept Buchanan and *Virginia* in Portsmouth, chaffing at the bit.

Monitor's crew spent a miserable time at the dock, the cold rain hammering down on the iron deck and the choppy water of the East River sweeping over the vessel. The ironclad's captain, John L. Worden, invited some of the shipyard's officers to dinner onboard the ship, but even that was ruined when the wardroom steward managed to get drunk by dipping into the champagne and brandy meant for the officer's table.

In the tiny ship, the anchor well was made to double as a brig, and it was there that the unhappy steward was placed in irons.

Mechanics from the navy yard were struggling to finish last-minute repairs, working around the clock, just as they were doing onboard *Virginia* in Portsmouth. Since the middle of February, the telegram wires between Washington, Fortress Monroe, and New York had been hot with messages asking when *Monitor* would be under way and with ever-changing orders for where she should go.

Finally, on March 6, the work was done and the weather moderate enough for *Monitor* to put to sea. At 10:30 P.M., the morning clear and cold, she slipped her dock lines and steamed down the river toward Sandy Hook and the Atlantic Ocean.

In company with the ironclad, hovering like watchful parents, were the wooden screw steamers *Currituck* and *Sachem*, both ex-merchant vessels purchased during the navy's frenzy of acquisition in the fall of 1861. Also part of the small convoy was the steam tug *Seth Low*.

At 4:00 P.M., with the weather still favorable, *Monitor* crossed the bar of New York Harbor and stood out into the open ocean, one of only two times in her career she would leave the coast astern. Worden dismissed the pilot and sent with him a note for Secretary of the Navy Gideon Welles, informing him of their having left New York harbor. "In order to reach Hampton Roads as speedily as possible," he wrote, "whilst the fine weather lasts, I have been taken in tow by the tug."

For the moment, it was fine weather indeed, particularly given the capricious nature of the North Atlantic at that time of year. The wind was light from the west and the seas so smooth that not a drop of water came over the deck, despite *Monitor*'s mere eighteen inches of freeboard.

In the wardroom, the officers enjoyed a convivial meal. Worden had suffered a relapse of the fever he contracted while a prisoner of war in Montgomery and was able to eat nothing but cod liver oil. But despite his ill health and limited diet he entertained the officers with tales from his midshipman days.

As the sun was going down, the ship's paymaster, William Keeler, climbed up on the top of the turret to enjoy the evening. The moon was shining down on the smooth water, and a smattering of sails could be seen along the horizon. Just to seaward, keeping their station, the *Mon-*

itor's escorts, the gunboats *Currituck* and *Sachem,* steamed along, their green starboard running lights gleaming in the gathering dark. It was a perfect evening, and Keeler was feeling buoyant when he finally went below.

The little convoy steamed on through the night, *Monitor* making five knots at the end of a 400-foot towing hawser, stretching from the aft bitts of the *Seth Low.* The ironclad's motion in the water was easy and stable, thanks to her wide beam and lack of any weight higher than the turret, 9 feet above the deck. Chief Engineer Alban Stimers, who had come along as a volunteer, claimed, "I never saw a vessel more buoyant or less shocked in a heavy sea . . . There has not been sufficient movement to disturb a wine glass setting on the table."

Even Samuel Greene, the twenty-two-year-old executive officer, who, after sailing in her, did not consider *Monitor* "a seagoing vessel," admitted that her "roll was very easy and slow, and not at all deep. She pitched very little, and with no strain whatever. She is buoyant, but not very lively."

By 3:00 A.M. the four vessels had covered ninety miles, with Absecon Light and Atlantic City off their starboard quarters. The two gunboats maintained their station to seaward of *Monitor*. The weather was still clear and cold, with moderate seas.

Then things started to get rough.

By dawn on the morning of the 7th, the wind had veered around to the northwest and strengthened and the seas were building. Keeler woke to the feel of the vessel working harder in the seaway. Through his deck light he could see green water rushing over the deck above, as *Monitor* dipped her low sides into the waves.

It is axiomatic at sea that one never knows where the ship's problems are until the first rough weather hits, and that principle was amply demonstrated onboard *Monitor*. Once the seas began breaking over the flat deck, the ship started leaking in a dozen places. Water came in around the deck lights and poured down around the imperfectly sealed hatches. Worse, it rushed in all around the base of the turret, a leak 63 feet long.

John Ericsson, the brilliant engineer who created *Monitor*, had designed the turret to rest on a bronze ring on the deck. The weight of the turret on the smooth, soft metal was intended to form a watertight seal.

But the sailors at the navy yard did not believe that the joint would be watertight. They resorted to the shipwright's ancient method of sealing leaks—inserting old rope as caulking.

In this instance, a plaited, or braided, rope was laid around the circumference of the turret, between the turret and the bronze ring. "As might have been supposed," Ericsson wrote later, "the rough and uneven hemp rope did not form a perfect joint." The frigid sea came in "like a water fall" and flooded out the berthing deck, directly under the turret, and poured into the engine room aft. All hands were soaked and miserable, and some, including Worden and the surgeon, Daniel Logue, were seasick.

With the seas rushing over the deck, the roof of the turret was the only place topside one could go. The seasick men retreated there for fresh air, one of the best remedies for *mal de mer*. Off the port side, Keeler observed the two gunboats still on station but making heavy weather of it, rolling hard, much harder than the stable *Monitor*. The ironclad had become "wet & very disagreeable," but she was not yet in danger.

Hell and High Water

All morning long the wind continued to build to gale force and the seas continued to mount. Where earlier the deck was just awash, now the waves were breaking over the ship as if she were a rock on the shore. Only the 9-foot-tall turret stayed above the surface as the seas made a clean sweep fore and aft. Cresting waves hit the pilothouse and burst into the air, arching "over the Tower in most beautiful curves." The water slammed into the sides of the pilothouse and shot through the eye slots with such force that it spun the helmsman around.

As the waves built, the sea found *Monitor*'s greatest Achilles heel. Just behind the turret were two square holes in the deck that served as smokestacks and, behind those, two more for air intakes that supplied the forced-air ventilation for the living spaces and the draft for the boiler fires. Like everything on *Monitor*'s deck, they were designed to be flush with the deck when the vessel was cleared for action. Realizing that there was a chance that water would come over the deck, Ericsson had provided square stacks that could be fitted over the openings. The smokestacks were six feet high, the stacks on the air intakes four.

But Ericsson was not a sailor and did not comprehend how high the seas could get. Worden did, and he objected to the size of the stacks, telling Ericsson they were too short. Ericsson disagreed and resented Worden's objections, just as he resented any insinuation that anything he did was less than perfect. At last Worden gave in, telling Ericsson, "You build your vessel and I will sail her."

But Worden was not the only one who recognized the problem. Alban Stimers, now a passenger onboard the ironclad, was the navy's superintendent of construction for the *Monitor* while she was being built. He was a skilled engineer, a friend and follower of John Ericsson, and he also spotted potential trouble. "The cause of the difficulty was that the air pipes that were to keep the water out of our blowers were not high enough," he wrote. "If Captain Ericsson had made them as high as I wanted them (same height as the turret) we would have suffered little inconvenience."

Stimers brought this to Ericsson's attention many times, but Ericsson would not listen. "[H]e was very obstinate," Stimers wrote, "and insisted that four feet was high enough."

It was not. Water poured down the stacks, setting off a chain reaction of failures. Directly under the air intake openings in the deck were blowers. The blowers forced air into the fires in the boilers driving the engine and provided fresh air for the entire belowdeck ventilation system. The water soaked the belts that drove the blowers, causing them to slip, then stretch and break. As the blowers failed, the fires in the boilers began to die for lack of air. As the fires went out, the steam pressure dropped off—the very pressure that was needed to drive the bilge pumps that kept the ship from sinking.

Water also cascaded down the smokestacks and vaporized in the intense heat of the fireboxes. The blasts of steam in the fireboxes forced poisonous gases, "hydrogen and carbonic acid," out through the furnace and ash pan doors. Steam, smoke, and toxic gas filled the engine room, where the engineer's division, known as the black gang, was struggling to fix the blowers. The engine room could not be cleared of the gas without the blowers to force it out.

Monitor had two blowers. When the belt on the first one broke, the ship's engineer, Isaac Newton, leapt to repair it before the air in the engine room became untenable. He had almost managed to get a new belt in place when the second blower gave out.

Newton knew he needed help. He reported to Alban Stimers, and Stimers raced to the engine room. With the assistance of the other engineers and firemen, Stimers concentrated on the first blower which had a new belt nearly in place. But it was no good, he could not get the blower operating, and the space was filling with toxic gas. One by one the men began to collapse at their stations, overcome. Stimers ordered everyone out of the engine room.

Alerted to the situation, Lieutenant Greene led a gang of seamen aft to the engine room, where he was greeted by a shocking scene, the engine room choked with gas, smoke, and steam, the engineers, firemen, and coal passers collapsing in heaps on the iron deck plates, "apparently as dead as men ever were."

Greene led his crew into the hellish place and began dragging the unconscious black gang out through the oval door from the engine room to the berthing deck. Soon the sailors, too, were gagging, choking, their heads swimming, as they struggled to get the dying engineers to fresh air.

Stimers remained in the engine room. He had to get just one blower online. If he could do that, then *Monitor* would live.

But he could not. Try as he might, he could not get the blower to work, and the gas was starting to overwhelm him, too. His knees were weak and he could no longer think straight. Finally, on the verge of collapse, he staggered through the bulkhead door to the berthing deck.

William Keeler was up on the turret, watching, transfixed, as *Monitor* worked in the big seas, rolling her lee side under, then coming up again, shedding water "like Niagara Falls." Green water rolled over the deck, breaking against the pilothouse and turret, completely submerging the hull so it looked as if the ship would be buried by the sea. The gunboats were still on station, rolling so hard that they dipped the muzzles of their guns in the water. The four vessels were near the Delaware/Maryland border, ten miles offshore. Fenwick Island Light—the starting point of the Mason-Dixon Line—was bearing west by south.

Around four o'clock in the afternoon, Keeler decided to go below.

As he climbed down through the small hatch he met an engineer coming up, "pale, black, wet & staggering along gasping for breath." The engineer asked Keeler for brandy, and Keeler agreed to get him some. The paymaster dove below, right into the nightmare on the lower deck.

While the solid iron door in the bulkhead between the engine room and berthing deck was closed, the smoke and gas was confined to the after part of the ship. Once the black gang began to evacuate and the door was flung open, the entire ship was filled with the deadly fumes. Keeler went below just as Greene's men were hauling the last of the engineering department out of the engine room and up the turret hatch. By now the sailors were "stifled with the gas," nearly as bad off as the engineers.

Keeler could see that the smoke and gas were billowing out of the engine room door, which was still hanging open. Choking and suffocating, he staggered over to the door to shut it, but one of the sailors still below told him that there might be one man left behind.

Keeler charged into the engine room, stepping over heaps of coal and ash, and found the man, unconscious and lying on the floor plates. Keeler and a sailor who had followed him in dragged the engineer out and shut the engine room door. They wrestled the half-choked man up the ladder to the turret, and then up another ladder to the roof of the turret, the only place now where the crew could find breathable air. By the time they pulled him topside, the engineer was nearly dead.

Night was coming on and all of *Monitor*'s crew was huddled on top of the turret. An umbrellalike awning covered the turret and a piece of canvas was rigged to windward across the awning stanchions to minimize the spray coming onboard. The ensign was run up the staff inverted, a universal distress signal, in hopes that the *Currituck* and *Sachem* would see them. But the gunboats were fighting their own battle and would have been of little use in that wild sea.

As Greene put it, "times looked rather blue . . ."

The Old Tank

Stimers had just managed to make it to the roof of the turret, where he found "three engineers and several firemen senseless." He sprawled out flat, gasping for air.

Soon after, Isaac Newton recovered enough to regain his sense of duty, "though not enough to use his best judgement." Newton decided that he had to go back to the engine room. Stimers, though he outranked Newton, had no official position onboard *Monitor* and could not stop him. Newton lunged into the smoke and gas again, rigging a hose to wet down the fires, which would presumably quench them and put a stop to the deadly gas.

He got no further than attaching the nozzle to the hose before he was once again overcome. He staggered to the engine room door and made it onto the berthing deck before he collapsed. Fortunately, a fireman saw him go down and dragged him topside again. For fifteen minutes the surgeon worked on him, until he slowly began to revive, though "his case looked doubtful for several hours."

With the blowers offline, the fires in the boilers' fireboxes began to die and the steam pressure dropped off. The turns on the propellor slowed to a near stop and the pumps would not work.

Lieutenant Greene led a gang of men below. With the fires going out, the smoke and gas had likely dissipated, as the sailors were able to remain below while they tried to find some way of keeping the ship afloat.

The water was deep and rising fast. Greene ordered the hand pumps rigged on the berthing deck. There were only two scuttles through which the discharge hoses could be passed. One was in the deck, forward, which clearly could not be opened. The other was through the turret, more than 17 feet above the berthing deck.

The hoses were run up through the turret, but the hand pumps could not generate enough force to lift the water that high. The only option left was to bail, but that was pointless. It would take far too long to hand the buckets up through the turret to be dumped over the side. There was nothing more they could do.

The only thing in *Monitor*'s favor was the direction of the wind, which was offshore, allowing them to approach the coastline with less fear of being driven onto the beach. By late afternoon, they managed to hail the tug, 400 feet ahead, and signal to it to tow the *Monitor* closer in to the land, where they hoped to find calmer water.

For several hours the *Seth Low* towed the disabled ironclad toward the Maryland shore, hard steaming for the small tug. During that time Stimers, on the roof of the turret, dispatched men in intervals to race down to the engine room and perform a given task, then get topside again before they passed out. With all of the engineers disabled, he had to rely on the firemen to perform the engineering duties, but in that way he managed to make progress in setting the blowers to rights.

It was around 8:00 P.M. when the *Seth Low* finally towed the *Monitor* into calmer water near shore and the engineers could go back in the engine room. Able to breathe, and with no more water coming down the blower pipes, they soon had the fires going again. Steam was applied to the engine and pumps and the ship was freed of the water she had taken on. With the crisis seemingly over and the systems back online, the *Monitor* was under way once more.

By the time the ventilators had blown all of the toxic fumes out from the ship's interior, it was full dark. The men stumbled below and dined on crackers and cheese and water. *Monitor* continued on her course, south-southwest, making five knots through now calm seas.

Captain Worden had had little sleep in the past thirty-six hours, and that, combined with the vestiges of fever and the diet of cod liver oil, left him nearly dead on his feet. Lieutenant Greene offered to take the watch from eight o'clock to midnight and Worden agreed, retiring to his cabin for a blessed four hours' sleep.

Greene passed the watch on the *Monitor*'s turret, steaming along through ideal conditions with "smooth sea, clear sky, the moon out and the old tank going along fine and six knots very nicely." When the change of watch came up at midnight, Greene felt that conditions were so good that there was no need for Worden to turn out. Greene offered to let Worden sleep, the executive officer agreeing to lay down with his clothes on—all standing, as it is called—so that he could be called out instantly if there was a problem. Worden agreed, and the men went to their respective cabins to rest.

No sooner had he laid down than Greene was rocketed out of bed by the most "dismal, awful sound" he had ever heard, a sound that would make the din of battle sound like music. It resembled the "death groans of twenty men."

Greene rushed from his cabin and Worden did as well. *Monitor* had passed over shoal water and once again found herself in steep, rough seas, this time coming bow-on. In the forward end of *Monitor*'s deck there was a round anchor well, the top of which was covered with two-inch plate iron. As the bow slammed down into the head seas, the air in the anchor well was compressed. Its only outlet was through the hawse pipe, which ran from the after end of the anchor well to the interior of the ship and allowed the anchor chain to come onboard. The effect was like blowing a giant horn right into the wardroom.

Since the hawse pipe was mere inches above the waterline, more than just air was coming in. With every wave that came up under the bow, the seawater came jetting in "in a perfect stream." Greene claimed that the water came in with such force that it ran over the wardroom table, 30 feet away, before flowing aft to the berth deck.

Worden and Greene struggled in the freezing water to plug the hawse pipe (Ericsson claimed it was a "gross oversight on the part of the executive officer" that it was not plugged before they left the harbor, and he had a point). They managed to get the pipe at least partially blocked, but by then the water was coming down the blower pipes again, threatening the blowers, as the nightmare of that afternoon repeated itself.

From the top of the turret they tried to hail the *Seth Low*, but being downwind of the tug they could not be heard, nor could they be seen in the dark. It was a dismal, bleak moment, and the exhausted men began to despair of living to see the sunrise.

Once again the seas were making a clean sweep over *Monitor*'s decks. Anxiously, Greene sent for continuous reports on the state of the blowers. Every time the firemen answered the same—the blowers were going slowly, but they could not be kept going much longer unless the water stopped coming down the pipes.

Around one o'clock in the morning, with the *Monitor* pitching hard into the head seas, the tiller ropes came off the steering wheel and jammed themselves into a great tangled mess. As the *Seth Low* blithely towed the ship along, *Monitor* began to veer wildly side to side, out of control. The officers felt certain the hawser would part under that incredible strain, and then *Monitor,* her engine barely turning, her steering gear disabled, would be at the mercy of the sea. But the hawser was new, and it held.

The sailors struggled with the tiller ropes, but they could not get them rigged again. Finally they turned out Alban Stimers, who managed to get the ropes back on the wheel and working. The helmsman now had control of the rudder and was able to keep the vessel from sheering off, but the seas were still breaking across the deck and pouring down the blower pipes.

By 3:00 A.M. the seas had gone down a bit, though significant water was still coming onboard. Greene continued to ask for reports. "The never failing answer from the Engine Room," he wrote, "'Blowers going slowly, but can't go much longer.'"

It was a bleak time onboard the ironclad as the men once again found themselves in great jeopardy with nothing they could do. This time their hopelessness was no doubt augmented by the dark and their own exhaustion. Samuel Greene described the time from 4:00 A.M. until dawn as "certainly the longest hour and a half I have ever spent." He felt as if the sun had stopped in China and would never come back to their stretch of the North Atlantic.

Hampton Roads

Greene's fears notwithstanding, the sun finally broke the horizon to the east, and the men on *Monitor*'s turret were able to signal the *Seth Low* that they needed to head more inshore. For the second time in ten hours the tug pulled the ironclad into calmer waters. By eight o'clock in the morning the danger was passed, though the seas continued to roll over *Monitor*'s deck, coming in around the tower and hatches and preserving the high level of misery felt by the men below.

By 1:00 P.M. they were once again in "fine clear weather" just 11 miles from the entrance to the Chesapeake Bay with Smith Island lights, and Cape Charles, bearing southwest by west. At two-thirty the much-abused hawser finally parted, but with the weather moderated they had no difficulty in bending on a fresh one.

Around 4:00 P.M. the little convoy was steaming past Cape Henry, the southern point marking the entrance to the bay and the route to Hampton Roads. Far off, they could hear the muted sound of gunfire. Gray clouds of gun smoke hung over the low shoreline in the distance. Something was happening. They did not know what, but they could guess.

Worden ordered the *Monitor* cleared for action. She was "stripped of her sea rig," meaning that the pipes on the smokestack and blower intakes were removed, iron deadlights placed over the deck lights, and the awning removed from the top of the turret. The turret was keyed up so that it could revolve freely and the useless plaited rope removed. The massive iron round shot was hoisted up into the turret.

As the *Monitor* steamed closer to Hampton Roads the men could now see the streak and burst of shells in the air. The sound of gunfire was louder. Curiosity, anxiety, and fear mounted.

By 7:00 P.M. the *Monitor* was 12 miles from Fortress Monroe. The flash of bursting shells was vivid in the fading light. A pilot boat came out to meet the arriving vessel.

The pilot who came onboard was full of news. He described the devastation that *Virginia* had wrought that day. He told the stunned men about the sinking of the *Cumberland*, the *Congress* set on fire, the other ships awaiting their destruction at the hands of the invulnerable Confederate ironclad. A strange mood settled over the ship—anger, sadness, a desire for vengeance, a desperate longing to get into the fight.

The *Monitor*'s crew was exhausted. They had come about as close to sinking as they could come, but they had lived. They were fifty-one hours out of New York City, and they were six hours too late.

Chapter 3

Birth of the Ironclads

Like the stars in the wake of the Big Bang, the CSS *Virginia* and the USS *Monitor* were both born of the same cataclysmic event, the destruction of the Gosport Naval Shipyard.

There were plenty of people to blame for that debacle, but the chief fault rests with Abraham Lincoln and his secretary of the navy, Gideon Welles. And while Welles would grow into his office, becoming ultimately an excellent secretary of the navy, his start was inauspicious at best.

Jefferson Davis was fortunate in Stephen Mallory, the man he selected to run the Confederate navy. Just as fortunate, perhaps more fortunate, despite the early failures, was Lincoln in his selection of Welles.

Like Mallory, Welles was shrewd and competent. Like Mallory, he would serve his president well for the duration of the war.

Unlike Mallory, Welles would oversee a large and vital department. While the Confederate navy was never able to do more than offer minor assistance in staving off defeat, the Union navy under Gideon Welles would play a critical role in the eventual Union victory.

Gideon Welles did not particularly want to be secretary of the navy. His first choice was postmaster general, but that office went to a close friend of his, the influential Montgomery Blair from Maryland, whose state was teetering between the Union and the Confederacy.

Welles was fifty-nine years old in 1861, but looked older than that, with his thick white beard, wavy salt-and-pepper hair, and hawklike face. He looked like the very model of the taciturn, hard-driving Yankee sea captain, which he was not.

Gideon Welles was not a man of the sea, but rather a man of politics. He was born in Glastonbury, Connecticut, in 1802 to a fairly prosperous and influential family. The Welles line in Connecticut dates back to around 1633, when Thomas Welles, who would serve as governor, emigrated to that colony.

Gideon Welles studied law but did not practice it. He was drawn instead to the political area, an ardent Jacksonian Democrat. He became editor and co-owner of the *Hartford Times and Weekly Advertizer*. He served eight years in Connecticut's state legislature and was the state's comptroller and postmaster of Hartford.

A rising star in the Democratic Party, Welles was given the post of chief of the navy's Bureau of Provision and Clothing in the administration of James Polk, making Welles the first civilian to serve as a Navy Department bureau chief. The job he had wanted, and was destined never to get, was that of postmaster general, and he was not happy about the considerably less prestigious position he was given. Still, he approached his duties with honesty, intelligence, and drive, and brought to the bureau an efficiency and organization that it had not hitherto known.

By the time of the presidential elections of 1860, Welles had left the Democratic Party in exasperation over its policy of tolerating the expansion of slavery and joined with the newly formed Republicans. Welles, now a well-connected and influential political operative, was named to the Republican Party's executive committee. In that capacity he first met Abraham Lincoln in March 1860, a little more than a year before the two of them would together face the prospect of going to war with their fellow Americans.

When Lincoln managed to win the presidency with only 40 percent of the popular vote, Gideon Welles was one of the political operatives he had to thank, and Welles's many supporters felt a cabinet position would be thanks enough. With the office of postmaster general going to

Blair, Welles was offered the navy. Like a character straight out of
Gilbert and Sullivan's *H.M.S. Pinafore*, Welles's hard work and party
loyalty—and his never going to sea—took him from an apprenticeship
in a law firm to the leadership of Lincoln's navy.

Political appointments are often hit-or-miss propositions, and
Abraham Lincoln suffered a number of misses. Generals and admirals proved incompetent, cabinet members were schemers or crimi-
nals or both. Until Lincoln reigned him in, Secretary of State
William Seward was quite active in undermining the president's authority.

But in Gideon Welles, Lincoln got just the man he needed. Welles
was intelligent and calm in the face of crisis, with enough imagination
to see the potential in new ideas, where tradition-bound naval officers
often dug in their heels. He had no patience for incompetence and did
not believe that seniority was the most important quality that a fighting
sailor should have. And though Lincoln privately referred to Welles as
"Father Neptune" and "Uncle Gideon," he came to appreciate the dour
Yankee's abilities.

Mr. Welles Goes to Washington

Like most in the Lincoln administration, Welles stepped straight from
private life into public crisis. And while the inevitability of civil war
seems clear in hindsight, it was not nearly so obvious in the confused
spring of 1861. Welles neatly describes the prevailing spirit he encountered in the nation's capital:

> A strange state of things existed at that time in Washington. The at-
> mosphere was thick with treason. Party spirit and old party differ-
> ences prevailed, however, amidst these accumulating dangers.
> Secession was considered by most persons as a political party ques-
> tion, and not as rebellion. Neither party appeared to be apprehensive
> of or to realize the gathering storm. There was a general belief in-
> dulged in by most persons, that an adjustment would in some way be
> brought about with out any extensive resort to extreme measures.

Clearly, Lincoln was one of those who hoped to avoid "extreme measures." Initially, the Lincoln administration's approach to the crisis was the same as the classic advice to physicians: first, do no harm.

The situation was bad enough, with seven states seceded and a Confederate government established in Montgomery, Alabama. But it could still get worse. Virginia, foremost of the Southern states, had not yet seceded. Nor had Maryland. Lincoln wished to keep it that way. As Welles wrote, "That there should be no cause of offence, no step that could precipitate or justify secession, the President, almost daily, enjoined forbearance from all unnecessary exercise of political party authority." Lincoln did not want to give either state reason to leave the Union.

Welles felt differently. "A heavy hand," he wrote, "could it have been placed on these wretches who advocated treason and urged disunion . . . would have been better than attempts at reconciliation."

Unlike Secretary Seward, Welles did not take it upon himself to craft policy. Still, he understood the precarious nature of things. If they made an attempt to defend federal interests in Virginia, it could inflame the secessionists and tip the state toward the Confederacy. If they did nothing, it would leave valuable assets vulnerable to capture by the enemy.

The question of how boldly to defend U.S. government installations was a sticky one. The entire standoff in Charleston, after all, revolved around the question of who had the right to federal property—the United States or the state in which the property resided. Lincoln already had one Sumter on his hands; he did not want another.

For the Navy Department, the greatest and most threatened asset was the Gosport Naval Shipyard.

Though the shipyard was generally referred to as the Norfolk Shipyard, it was, and still is, actually located across the Elizabeth River from Norfolk in the town of Portsmouth. Since the name Portsmouth Naval Shipyard was already taken by the shipyard in Portsmouth, New Hampshire, the yard in Virginia was named for the nearby town of Gosport.

Whatever the history of its name, the shipyard was an old and invaluable resource. The first yard on that site had been established in 1767. By 1793 it was officially designated the Gosport Naval Shipyard, one of the largest and most productive in the United States.

By the time Lincoln took office, the shipyard boasted one of only two dry docks in the country (the other was in Charlestown, Massachusetts), the largest shears, or cranes, in the United States, machine shops, boiler shops, carpentry shops, rope walks, timber sheds, mast houses, burnetizing houses in which canvas, wood, or rope was treated with a solution of chloride and zinc, and two towering ship houses, with another under construction. Along with the land-based assets, there were a number of valuable men-of-war docked there in various states of repair.

Lincoln ordered Welles to make no extensive changes of personnel or readiness to the navy yard—nothing that could be construed as hostile or defensive—without consulting him. Lincoln felt that "any extraordinary efforts to repair the ships, with a view to removing them and the public property would . . . exhibit a want of confidence and betray apprehensions that should be avoided." Welles agreed, but soon found himself growing increasingly concerned for the yard's safety.

On March 14, the storm-damaged *Pocahontas*, a screw steamer of the Home Squadron, arrived in Hampton Roads from Vera Cruz. Welles dispatched the ship, with her four 32-pounder guns, a 10-inch rifle, and a 20-pound Parrott rifle, to the Gosport Naval Shipyard, ostensibly for repairs, but also to aid in the yard's defense. He was confident that the secessionists would see nothing unusual or threatening about a man-of-war being sent to a naval shipyard. Likewise, when the frigate *Cumberland* returned from the West Indies on March 23, Welles sent her to Norfolk as well. But he was still not happy with the situation.

Several times Welles applied to General-in-Chief Winfield Scott for troops to be sent to the yard, but each time Scott assured him there were hardly troops enough to garrison Fortress Monroe* and Harper's Ferry. This was certainly the case, as the army was small to begin with, and a good deal of it was stationed in the far West. Besides, Scott assured Welles, Norfolk was wholly indefensible in any case.

Exasperated, Welles appealed to President Lincoln, but Lincoln sided with Scott. As Welles explained it:

---

*Monroe, the huge army installation that even today overlooks the entrance to Hampton Roads, was known alternately as "Fort Monroe" or "Fortress Monroe," apparently based on the writer's preference. This important and powerful base formed the Union's main foothold in the Hampton Roads area.

"[T]he President thought it would be inexpedient and would tend to irritate and promote a conflict were a military force to be sent to Norfolk."

Alternate Plans

By the end of March, Welles might well have considered the shipyard a lost cause. Things in Charleston were growing more tense by the day, and the administration was torn as to whether or not to attempt relief of Fort Sumter. What's more, Welles had never really believed that Virginia would remain in the Union, feeling that the state was in the hands of men grown rich on the slave trade who would never align themselves with Lincoln.

Since he was receiving no help in defending the yard itself, and realizing that it might be beyond saving, Welles began to think about the ships.

Several of the ships at Norfolk were relics of a bygone era, such as the old sailing line-of-battle ships *Pennsylvania* and *Delaware*. Some were sailing vessels but still valuable, such as *Plymouth, Germantown*, and of course *Cumberland*, which Welles had recently sent to the yard. But the most valuable ship by far—worth by some estimations as much as all the others combined—was the steam frigate USS *Merrimack*.

Merrimack was only five years old in 1861. Thirty-two hundred tons, 275 feet long, 38 feet, 6 inches on the beam, drawing 27 feet 6 inches, she was one of the most powerful warships in the U.S. Navy. She was one of six heavy screw frigates built in 1856, the first ships in the United States navy to be driven by propeller rather than the vulnerable side wheels of an earlier generation. The others were *Niagara, Roanoke, Colorado, Minnesota*, and *Wabash*, all of which would play their roles in the coming war.

For armament she carried fourteen 8-inch guns, two 10-inch guns, and twenty-four 9-inch Dahlgren smoothbores. She had last served as flagship of the Pacific squadron. If *Merrimack* had one defect, it was her two cylinder double-piston-rod, horizontal, condensing engine, which was notoriously unreliable and inefficient. As a consequence, the boilers burned an inordinate amount of coal.

In March 1861, *Merrimack* was docked at the Gosport Naval Shipyard and all but disassembled. Her guns had been taken ashore, her masts, yards, and rigging stripped down to lower masts, and her engines and boilers pulled apart, their pieces scattered over the yard.

On March 31, Welles began to take steps to get the ships, particularly the *Merrimack,* to safety. He sent a private order to Commodore Samuel L. Breese, commanding the Navy Yard at New York, to ready 250 "seamen, ordinary seamen, and landsmen, to be transferred from the receiving ship *North Carolina* to the receiving ship *Pennsylvania* at Norfolk." Ostensibly the men were for the ships *Plymouth* and *Germantown*. Both ships were ready to get under way, and it would therefore seem reasonable even to secessionists that sailors would be sent to man them. But really Welles wanted the seamen there in case he had to move *Merrimack* in a hurry.

At Gosport, Gideon Welles had problems beyond secessionists and dismantled ships. One of those problems was Commodore Charles S. McCauley, the feeble, alcoholic sixty-eight-year-old commander of the navy yard. Welles had not seen McCauley for several years, and he had no reason initially to suspect he was not competent to fulfill his increasingly difficult duties. Welles asked people who knew McCauley better than he if the old commodore was up to the task of managing the yard in the coming crisis. He was told that McCauley could handle the situation.

But as things began to heat up, Welles's doubts began to grow. When Welles wrote McCauley asking how long it would take to get the *Merrimack*'s engines working again and McCauley replied that it would take at least a month, Welles became suspicious.

There was no thought that McCauley had Southern sympathies. His loyalty to the Union was never in question. As Welles later wrote, "subsequent events proved him faithful but feeble and incompetent for the crisis. His energy and decision had left him, and whatever skill or ability he may have had in earlier years in regular routine duty, he proved unequal in almost every respect to the present occasion."

McCauley's chief problem, besides incompetence and drink, was that a number of his subordinate officers were Southern sympathizers with plans to preserve the yard and the ships for the Confederate cause. One witness to the entire affair reported later, "The old hero

[McCauley] . . . could have none but the best and purest motives in all he did, but he was surrounded by *masked traitors* whom he did not suspect, and in whose advice he thought there was safety."

Among those masked traitors were men in very high positions in the shipyard who would go on to serve the Confederate States Navy with distinction, including John Tucker, the navy yard's ordnance officer, and the yard's executive officer, Robert Robb (who, more than a year later, would sit on the court-martial of Josiah Tattnall for the loss of CSS *Virginia*, the second time the old *Merrimack* was sunk).

Throughout the first week of April, Welles and Lincoln continued to equivocate regarding the disposition of ships and equipment at the Gosport Naval shipyard. Their focus, and that of much of the administration, was on the standoff at Fort Sumter and a similar situation at Fort Pickens in Pensacola, Florida. Supplies at Sumter were running low, and efforts to resupply might be construed as acts of war. The tension on both sides was reaching the breaking point.

On April 10, Welles wrote confidential instructions to McCauley that "the steamer *Merrimack* should be in condition to proceed to Philadelphia or to any other yard, should it be deemed necessary, or, in case of danger from unlawful attempts to take possession of her, that she may be placed beyond their reach." But once again, Welles tempered his orders with the familiar warning "that there should be no steps taken to give needless alarm." That kind of uncertainty did nothing to help the already uncertain McCauley, but was undoubtedly a benefit to the Southern sympathizers under his command.

The next day, Welles seemed to come to the conclusion that bolder steps were needed, no doubt prompted by the escalating crisis in Charleston. He ordered Commander James Alden to proceed immediately to the Gosport Naval Shipyard. Alden was given two sets of orders. The first, to be placed on file, was simply for him to report to McCauley. The second, written by the chief clerk himself and highly confidential, instructed Alden to assume command of the USS *Merrimack* and get her to the naval shipyard in Philadelphia.

Welles was making an extra effort to keep secret orders secret. With so many people of divided loyalties still in government, official business, it seemed, all too quickly became everyone's business. As it turned out, Alden's orders were no exception, and Welles's precautions did little good.

When the commander arrived in Norfolk he came ashore with the distinct impression that everyone in town knew why he was there. The people seemed to watch his every move, and follow him from place to place. Finally, Alden destroyed his written orders and moved from the comfort of his hotel to the security of the shipyard to wait on *Merrimack*'s repair.

On April 11, 1861, Alden was ordered to prepare *Merrimack* for sea. On April 12, Confederate forces in Charleston Harbor fired on Fort Sumter, and everything changed.

On that same day, and before the administration knew for certain what had taken place in South Carolina, Welles took the next step to get *Merrimack* to safety. Believing that McCauley must be wrong in thinking it would take a month to get *Merrimack*'s engine repaired, Welles asked the opinion of the navy's top engineer, Engineer in Chief Benjamin Franklin Isherwood.

Isherwood replied that in his opinion it would take perhaps a week to get the engine turning. Eager to get the ship to safety, Welles dispatched Isherwood himself to Norfolk to personally see the power plant made ready.

Isherwood arrived in Portsmouth on Sunday, April 14. He carried with him orders from Welles to McCauley, stating: "The Department desires to have the *Merrimack* removed from the Norfolk to the Philadelphia navy yard with the utmost dispatch. The engineer in chief, B. F. Isherwood, has been ordered to report to you for the purpose of expediting the duty, and you will have his suggestions for that end carried promptly into effect."

After meeting with McCauley, Isherwood conferred with the yard's chief engineer, Robert Danby. Together they inspected *Merrimack*'s engines, and what they found was not an inspiring sight. The braces were entirely removed from the boilers, the air pumps completely disassembled, and the engines "greatly out of repair."

Isherwood and Danby then conferred with the foremen of the various shops that would be involved in putting the engines and boilers back together. The foremen were instructed to gather every available workman—no easy task with so many of the yard's native Virginians abandoning their jobs—and prepare them for working two shifts around the clock.

The next day, Monday the 15th, deep in the *Merrimack*'s hull, work on the engines began. Outside, in Portsmouth and Norfolk and Washington, the country was moving rapidly toward war.

On the same day that Isherwood and Danby went to work, Abraham Lincoln took his first real bold, if naive, step toward military intervention by calling for the enlistment of 75,000 militiamen for ninety days' service, primarily to "re-possess the forts, places and property which have been seized from the Union" and to "redress wrongs already long enough endured."

Those border states still wavering between North and South chose to see this as an unwarranted act of aggression against their fellow Southerners. They informed the White House that no men would be forthcoming. The call for troops galvanized the South in the same way that the firing on Sumter had inflamed the North. The time for moderation, it seemed, was over.

But Virginia had yet to secede, and both sides wanted her. Virginia's decision to go Union or Confederate would have major implications for the coming war. Virginia had the largest population of all the Southern states, and nearly as much manufacturing capacity as all of the original seven seceding states combined. Richmond was home to the Tredegar Iron Works, the only plant in the South capable of manufacturing heavy ordnance or producing iron plating in quantity. The leading role that Virginia had always played in American history would lend legitimacy to any cause it backed. Virginia was the land of Washington and Jefferson and, of greater significance, as it would turn out, of Robert E. Lee.

Lincoln wanted Virginia. Welles wanted his ships. Welles did not think they would get both.

Chapter 4

Pawnee Steams South

On Tuesday, April 16, Gideon Welles sent Commander Hiram Paulding from Washington to Norfolk to verbally communicate the Navy Department's wishes, and to see for himself what was going on. Paulding was to act for Welles "in all particulars, provided danger was imminent, having plenary powers for the purpose." He was to confer with McCauley and with Commodore Garrett Pendergrast, who was there with the flagship of his squadron, the *Cumberland*.

Paulding carried orders for Pendergrast to put off his scheduled sailing for Vera Cruz in light of the difficulties brewing stateside. Paulding also carried more instructions for McCauley, which were unfortunately contradictory.

Welles ordered McCauley to waste no time in getting *Merrimack*'s armament onboard, as well as any other valuable property that might be saved from the shipyard. But then, having just days before told McCauley in no uncertain terms to send the *Merrimack* to Philadelphia, he now wrote that it "may not be necessary . . . that she should leave at that time, unless there is immediate danger pending." With McCauley already seized by uncertainty, such wavering on Welles's part did not help.

That night, Norfolk was lashed with a terrific rain and gusting wind, making the dark night even more impenetrable. The secessionists took advantage of the weather and made their first real effort to prevent the evacuation of the navy yard. Under the cover of the storm, two light-boats of about eight tons were towed to the narrow part of the river by Craney Island, where the Elizabeth empties into Hampton Roads. There the ships were sunk in an attempt to block the channel.

When the obstructions were discovered by men aboard the *Cumberland*, and the channel sounded, it was found that there was still sixteen feet of water, enough for the ships at Norfolk, including the stripped-down *Merrimack*, to pass. The lightboats were too small for the purpose, but they marked an escalation in hostilities that raised anxiety in the naval yard to a new level. McCauley ordered regular patrols of the yard, and started his handful of marines drilling with the howitzers.

Lieutenant Thomas Selfridge of *Cumberland* (who was so conspicuous in the battle with *Virginia*, and would end his years as Rear Admiral Selfridge) offered to man *Dolphin* with a crew of volunteers and anchor near the bottleneck at Craney Island, defending the river against further efforts to obstruct it. The plan was embraced by *Cumberland*'s captain, Captain Marston, and McCauley.

But before Selfridge could act on his idea, McCauley's subordinates talked him out of it, arguing that Selfridge, only twenty-four years old and just recently passed as lieutenant, might do something impulsive and cause an incident. McCauley changed his mind and the *Dolphin* remained in place.

Commodore Paulding arrived on Wednesday to find that Isherwood and Danby had performed a near-miracle. Even with her engines torn apart and scattered over the yard, and much of the work force gone, they had managed in less than seventy-two hours to get the *Merrimack* sufficiently repaired that she could raise steam and leave the yard under her own power.

The *Merrimack* was taking on coal when Paulding arrived and delivered Welles's instructions to McCauley and Pendergrast. Among the three men, it was decided that *Cumberland* should be moved to a place where her batteries could command the front gate and the western part of the yard. It was also agreed that once *Merrimack* had her coal onboard she would be moved under the shears to receive her guns, and then sent away.

Paulding returned to Washington the next day, much to Welles's surprise, who had intended for him to remain and keep an eye on things. To Welles's further surprise, Paulding reported that everything was fine at the naval yard. He told Welles that some of the young Southern naval officers stationed there, while still loyal to the Union, wished to be relieved of duty, as their continuing on at the yard made it awkward for their families who lived in town. He reported that McCauley was competent to handle the situation and that the public property was well protected.

The report was, in Welles's words, "more favorable than I had expected."

In retrospect, it is hard to imagine how Paulding came to the conclusions he did. But in fact, things had not yet reached crisis stage in Norfolk. It would be another day and a half before the real panic set in.

Indecision

With *Merrimack* coaled and ready to go, Isherwood and Danby went to meet with McCauley. It was the afternoon of Wednesday, the 17th. They reported that the ship was ready for steam, that there were forty-four coal heavers and firemen standing by in the engine room, along with three assistant engineers and Danby as chief. James Alden was there to command her with enough men and supplies to get the ship to safety.

Isherwood asked if they should fire up the boilers immediately, but McCauley hesitated, telling them that the next morning would be sufficient. Danby and Isherwood returned to *Merrimack* and divided the men into watches. Around daybreak the fires were started in the furnaces below *Merrimack*'s four huge boilers.

By 9:00 A.M., on April 18, the formerly disabled ship had steam up and her engine was turning at the dock. Once again Isherwood and Danby reported to McCauley. The only thing needed to get *Merrimack* under way was McCauley's permission, but again the old man hesitated, telling Isherwood that he had not yet decided whether or not to let the ship go.

Taken aback by this, Isherwood reminded McCauley that the orders from the Navy Department were quite explicit regarding the need to get *Merrimack* out of there. He pointed out that the ship could still pass over the obstructions sunk in the channel, but given more time the secessionists would add more obstructions and bottle the entire fleet in. But still McCauley would not give the order. Instead, he told Isherwood he would decide later that day. Unsure of what to do, and receiving conflicting advice, McCauley hoped to put off making a decision.

Leaving McCauley's office, Isherwood found Commander Alden and asked the officer to come aboard *Merrimack*. There he showed Alden the boilers with steam up and the engines turning and the engineering crew, the "black gang," all ready at their stations. "I told him [Alden] that so far as the engineer department was concerned the vessel was ready to go, and that my part was done." Isherwood had a sense for what might happen, and he wanted to cover his behind.

At 2:00 P.M., Isherwood once again called on McCauley. This time he found that the commodore had made a decision, but it was the wrong one. He was determined to keep *Merrimack* at the yard. Isherwood argued vehemently. He pointed out again that Welles's orders were "peremptory," that the obstructions in the river were of no consequence and the ship could safely steam into Hampton Roads and to Fortress Monroe, that they should send her immediately with the sloop *Germantown* in tow.

McCauley had been drinking, he had been listening to the advice of secret secessionists, and he would not budge. As Isherwood recalled, the commodore "sat in his office immovable, not knowing what to do. He was weak, vacillating, hesitating, and overwhelmed by the responsibilities of his position. . . . He behaved as though he were stupefied." McCauley ordered that the fires be drawn from the ship's boilers. *Merrimack* would not move.

Isherwood, disgusted, passed on McCauley's orders. "As I witnessed the gradual dying out of the revolutions of *Merrimac*'s engines at the dock," Isherwood wrote years later, "I was greatly tempted to cut the ropes that held her, and bring her out on my own responsibility." But he did not. He was a staff officer, not eligible for command of a vessel, and he was too much of a navy man to ignore that protocol.

Meanwhile, the people of Norfolk, incensed by the *Merrimack*'s repair, hatched a plot to capture Isherwood on his return to Washington.

A crowd gathered at the wharf to grab him as he boarded the Baltimore steamer. Fortunately for the engineer, a local man, a good friend of Isherwood's, discovered the plot. He booked a cabin aboard the steamer in his own name, and deposited Isherwood's trunk in it. Then he took a closed carriage to the hotel and retrieved Isherwood, taking him to the landing and slipping him unseen aboard the boat. It was only when the mob descended on the hotel that they realized Isherwood had escaped.

The engineer returned to Washington and reported to Welles.

Divided Loyalties

Isherwood was not the only one pushing McCauley to let the *Merrimack* go. Commander Alden was also trying to sway McCauley against the influence of the secessionist officers in the yard. How hard he tried varies, depending on whose version of events one reads.

According to Alden, he all but succeeded in getting his ship to safety. Meeting with McCauley, Alden came up against the same arguments that the engineers had: *Merrimack* was needed for defense of the yard, they could put her guns back aboard and use her as a battery, the obstructions in the river would prevent her from getting into Hampton Roads.

Alden argued strenuously. He pointed out that Lieutenant Murray of the *Cumberland* had sounded the channel and found plenty of water there. He mentioned the peremptory nature of Welles's orders. For support of his arguments, he appealed to Commodore Pendergrast and the captain of *Cumberland,* Captain Marston, who were in Mc-Cauley's office at the time. Pendergrast and Marston agreed with Alden, and under pressure from all three Union men, McCauley relented. *Merrimack* could go.

Commander Alden was eager to get under way before McCauley changed his mind again. He had a black gang, but he needed deckhands, since the men Welles had ordered from New York had not yet arrived. He asked Pendergrast for thirty men from *Cumberland,* to be returned once *Merrimack* was under the guns of Fortress Monroe. Pendergrast agreed and Marston left to make arrangements.

Despite Welles's order to McCauley to get *Merrimack*'s armament onboard, along with whatever equipment could be saved, nothing of the sort had been done. Orders to get *Merrimack* under the shears so that her heavy guns could be put back aboard were countermanded when it was suggested such a move would infuriate the secessionists in town and lead to an attack on the yard.

But Alden felt he needed some means of defending the ship. He asked McCauley for two field pieces that could be wheeled onboard without the use of the shears. McCauley told him to see the ordinance officer, Commander John Tucker.

Tucker, unfortunately, was a Southern man and would offer Alden no more than excuses as to why he could not have even a couple of light guns. Alden began to sense the secessionist forces acting under the surface, and that made him even more desperate to get his ship out of Portsmouth. He ordered Lieutenant Murray, a volunteer from *Cumberland*, to finalize negotiations with a pilot to bring *Merrimack* down the Elizabeth River. Alden told Murray to offer the pilot a thousand dollars for getting *Merrimack* to safety and another thousand if he could also bring out *Germantown*, as well as a place in the navy for life, which he would need, as he certainly would not be able to return to Norfolk.

Merrimack at this point was chained to the dock with her "head up steam" and her bow pointed up river. Alden ordered his first lieutenant to have the ship winded, or turned end for end at the dock, and have the dock lines singled. This would put *Merrimack* in a position to get under way in a matter of minutes once the men from *Cumberland* were aboard.

But the first lieutenant, rather than following those orders, informed Alden that first they needed permission from the yard's executive officer, Commander Robb. Alden hurried off to find Robb, meeting up with him just as Robb was leaving McCauley's office. Alden was too late. Robb, a secessionist, had worked his will on the faltering McCauley. Robb informed Alden that McCauley had changed his mind, that he, Alden, could not have the ship. The commodore had once again ordered *Merrimack*'s fires to be drawn.

Alden did not believe Robb, "whose loyalty I had begun to doubt." He stormed into McCauley's office, where he found McCauley "stupefied, bewildered and wholly unable to act." But Robb was correct:

McCauley had rescinded his order. *Merrimack* would stay. There was nothing more for Alden to do. Like Isherwood, he returned to Washington in disgust and reported to Gideon Welles.

It was Thursday, April 18. On that same day, Major Robert Anderson and the garrison from Fort Sumter arrived in New York City as heroes of the North. Robert E. Lee declined Lincoln's offer of command of all Union forces being raised to fight the insurrection.

And the day before, in secret session, the state of Virginia had voted to secede from the Union. All of Lincoln's appeasement had been for naught, and his attempts at caution would cost the Union a fortune in ships and guns. The Gosport Naval Shipyard was already lost to the United States government, and no one even knew it.

The Last-Ditch Effort

Gideon Welles heard first Isherwood, and then Alden, with growing alarm. And while Welles seems to have felt that Isherwood, whose only job had been to get *Merrimack's* engines running, had fulfilled his duty, he was very critical of Alden, believing that the commander's weakness had allowed the secessionist to railroad McCauley. Welles later wrote:

> Alden was timid but patriotic when there was no danger, for he was not endowed with great moral or physical courage, though he believed himself possessed of both, and was no doubt really anxious to do something without encountering enemies or taking upon himself much responsibility. At Norfolk all his heroic drawing room resolution and good intentions failed him. A man of energy and greater will and force, with the orders of the Secretary, would have inspired and influenced McCauley, whose heart was right, and carried out these orders.

Disgusted as he was with Alden, Welles knew the commander's story had to be heard. He took Alden to see Lincoln, who was meeting with his cabinet. After Alden told his tale and left, Welles suggested that the time had come to vigorously defend the yard and its valuable property. Lincoln and the cabinet agreed. Welles then reminded the president of Winfield Scott's ongoing reluctance to send troops to Nor-folk. Lincoln said he would talk to the old general himself.

Together, Welles and Lincoln walked from the White House to military headquarters to confer with Scott. But still the general-in-chief insisted that troops could not be sent, that he did not have the men, that the shipyard was the "enemy's country" and any troops would certainly be captured.

After much discussion, Scott relented, admitting that he had information that a battalion of volunteers from Massachusetts had recently arrived at Fortress Monroe, and that they could be sent to Portsmouth. At last, after a month and a half of walking a tightrope, Welles was able to take the action he thought appropriate. With Sumter captured by Confederate troops and Virginia's state government expressing its disdain for Lincoln's call to arms, there was no longer any concern over ruffling feathers. Welles began to frantically issue orders.

It was clear by now that McCauley was in over his head. Welles turned again to Commodore Hiram Paulding, ordering him to return to Portsmouth immediately. This time Paulding was not going there to observe, but to take command of the shipyard, and he was to bring with him soldiers, marines, and a man-of-war.

Welles fired off telegrams to the naval shipyards in New York and Philadelphia, and to the receiving ship in Baltimore, to send men to Norfolk. He told Samuel Breese, commander of the navy yard in New York, "Furnish the recruits that are sent to Norfolk with arms and ammunition. Let them be well armed." He ordered Commander Stephen Rowan of USS *Pawnee** to ready his ship for immediate departure, and he had one hundred additional marines assigned to the vessel.

Welles also understood that he might be too late to save the yard. In his written orders to Paulding, he said:

> *With the means placed at your command, you will do all in your*
> *power to protect and place beyond danger the vessels and property*

---

*For some reason, *Pawnee* was the ship most feared by the defenders in Norfolk, though she was not the most heavily armed. "The *Pawnee* was the 'bug-bear' . . . ," one soldier wrote. "Everybody had the '*Pawnee* fever.'"

belonging to the United States. On no account should the arms and
munitions be permitted to fall into the hands of insurrectionists, or
those who would wrest them from the custody of the Government;
and should it finally become necessary, you will, in order to prevent
that result, destroy the property.

To furnish Paulding with the means to destroy the yard, Welles sent
orders to Commander John Dahlgren, in charge of the ordnance department at the Washington Naval Yard, to load onboard the ship *Anacostia*
incendiary materials to be transferred to the *Pawnee*. Dahlgren gathered
together forty barrels of gunpowder, eleven tanks of turpentine, six
brushes, twelve barrels of cotton waste, and 181 paper fuses called port-
fires, enough to do some serious damage. Assisting him in that effort
was, ironically enough, Captain Franklin Buchanan, commodore of the
Washington Naval Yard, who would go on to command CSS *Virginia*,
but who had not yet resigned from the United States Navy.

Also accompanying Paulding was Commander James Alden, lately returned from the shipyard, as well as a number of officers assigned to take
command of the various ships at Norfolk and prepare them for removal.

Among the officers was the cantankerous Captain Charles Wilkes,
the explorer who had led the U.S. Surveying and Exploration Expedition to the Antarctic and who was the first to understand the Antarctic
to be a separate continent. Wilkes had a pretty low opinion of pretty
much everyone, later writing that Gideon Welles was "a fool," that
Paulding was "ignorant of his profession and dicta[to]rial in his manner," and that "[i]mbecility was the ruling power" in Lincoln's administration. Seven months later, Wilkes would bring the United States to
the brink of war with Great Britain when he removed the Southern
commissioners Mason and Slidell from the British ship *Trent*.

The army sent Captain Horatio Wright, an engineer charged with organizing the defense of the yard or, if need be, overseeing its destruction.

On the evening of Friday, April 19, Paulding left Washington
aboard the *Pawnee*. At two-thirty the next afternoon the ship arrived at
Fortress Monroe where they took on the Third Massachusetts Regiment, 360 raw recruits, some of the first to respond to Lincoln's call.

Around six in the evening, *Pawnee* was under way for Portsmouth. It was Paulding's plan to anchor the ships *Merrimack, Germantown, Plymouth, Dolphin,* and *Pennsylvania* in the river, where their guns would discourage attack and prevent the locals from sinking any further obstructions in the channel. Safe in the stream, Paulding could then begin to move the ships to safety in an orderly fashion.

He was four hours too late.

Chapter 5

Panic

Friday, the day *Pawnee* sailed from Washington, was the day the situation at the Gosport Naval Shipyard began to disintegrate.

Satisfied that the *Merrimack* was not leaving, and hearing word of Virginia's secession, nearly all of the naval officers under McCauley resigned their commissions and left to join the Confederacy. Of the 800 or so mechanics and laborers whom McCauley had once thought loyal, every one of them abandoned the yard.

With the spirit of rebellion sweeping Virginia, the people of Norfolk grew bolder in their designs on the naval facility. A small steamer from the custom house, carrying the surveyor of the Port of Norfolk and forty or fifty armed men, approached the yard and attempted to capture the navy tug *Yankee*. McCauley ordered his marines to defend the tug to the last. The Virginians abandoned their attempt in the face of the marines' muskets, and the standoff ended without bloodshed. It was, however, the first overt attack on the shipyard, and McCauley was certain that more, and worse, were to come.

That night, local militia staged a raid on the magazine at old Fort Norfolk. No resistance was encountered. The Virginians secured for themselves over 2,800 barrels of gunpowder for little more than the effort needed to haul it off.

All through that night and the following day, the besieged men in the navy yard could hear trains rumbling in and out of the station in town and the constant blast of train whistles. Local informants told them that troops were pouring in from all over Virginia. McCauley reported, "I understood that Virginia State troops were arriving at Portsmouth and Norfolk in numbers from Richmond, Petersburg, and the neighborhood, and not having the means at my disposal to get the *Merrimack, Germantown,* and *Plymouth* to a place of safety I determined on destroying them, being satisfied that with the small force under my command the yard was no longer tenable."

McCauley believed that by Friday the 19th, around two thousand men had arrived in Norfolk and Portsmouth and he was certain that the yard was about to be taken. He had already concluded that he must destroy the ships to prevent them from falling into enemy hands. But still he hesitated.

By the next day, McCauley's nerves were pushed beyond their limits. Rumors were circulating that Virginia troops were setting up batteries across the river and north of the shipyard that would command both the yard and the channel. Lookouts were sent aloft on *Cumberland* to try and determine the truth of that, but they could see no such activity. Still, there were enough wooded places to hide guns, and McCauley and Pendergrast were perfectly willing to believe the batteries were there.

Lieutenant Thomas Selfridge was sent to parlay with William Taliaferro, the general ordered from Richmond to take command of the local forces. Selfridge crossed the river to Norfolk under a flag of truce, which "fortunately was respected by the large and excited crowd collected on the waterfront." It must have taken some nerve to stand in front of that mob in the uniform of the U.S. navy.

Selfridge was taken to see Taliaferro. He told the general that if the Virginians continued to erect batteries, it would be considered an act of war to be acted upon accordingly. Taliaferro told Selfridge that it would be "a terrible thing to fire upon such an undefended city." Selfridge replied, "It would be a terrible thing if batteries were erected and the Yard and men-of-war present were destroyed."

The shipyard might have been vulnerable, but the *Cumberland* was not. With her twenty-two 9-inch smoothbore guns, a 10-inch smoothbore pivot gun, and a 70-pound rifle, the sloop-of-war could have destroyed Portsmouth and Norfolk at her leisure, and both sides knew it.

Taliaferro assured Selfridge that there were no batteries aimed at the shipyard. Taliaferro's aide-de-camp, Colonel Henry Heth, returned to the yard with Selfridge to personally assure McCauley that General Taliaferro was unaware of any batteries, and that if he did become aware of them, he would prevent their completion.

McCauley apparently accepted this as the truth, but as far as he was concerned, it was already too late. The yard was lost. Abandoned by the officers he had trusted, left without a workforce, threatened by thousands of armed men on the other side of the insubstantial 10-foot high brick wall that surrounded the shipyard, drunk and confused, McCauley was overcome by panic. He had no one to man the ships at the yard, no pilots to get them downriver, no means at all of getting the ships to safety. At four o'clock in the afternoon, he ordered the destruction to begin.

The First Sinking of the *Merrimack*

All the vessels that were potentially useful to the enemy, the *Germantown*, *Plymouth*, and *Dolphin*, were scuttled. The engines and machinery onboard *Merrimack* were broken up to the extent possible and then *Merrimack* was scuttled as well. The great shears were hacked down, and they fell across the sinking *Germantown* and crushed her under their weight.

Cumberland was spared the fate of her fellow ships. Despite being of less value than the larger steam frigate *Merrimack*, she was a vessel in active service, with a crew of 350 men, a captain, and a flag officer onboard who would not care to see their ship sacrificed.

Armed, equipped, fully manned, and anchored in the stream, *Cumberland* represented the only real force the federals could bring to bear against the Virginians, as well as being the only safe refuge for the beleaguered Union navy men. It was McCauley's intention, after seeing the ships scuttled, to defend the shipyard against attack, at least until the following day, or, if that was not possible, retire to the relative safety of *Cumberland* and there decide what he would do next.

Then, around eight o'clock that evening, *Pawnee* came steaming into sight, and McCauley was spared from having to make any further decisions.

At about the same time that the ships at Gosport were beginning to settle into the mud, *Pawnee*, jammed to the gunnels with sailors, soldiers, and marines, had passed Sewell's Point and began heading up the Elizabeth River. She met with a schooner, which hailed them and informed them of the ships sunk in the narrow part of the river. *Pawnee* skirted around the obstructions and stood on for Portsmouth with guns loaded and run out and the men at quarters. From her deck, *Pawnee's* crew could see crowds of men on the nearby shore, lantern light glinting off the barrels of their muskets. But the ship was not challenged.

Cumberland was also prepared for a night action. They had heard rumors that a steamer from Richmond, loaded with soldiers, was on its way for the purpose of attacking the flagship. Every evening *Cumberland* was cleared for action, men at the guns and boarding netting rigged. On the night of the 20th, around eight o'clock, the drums suddenly beat to quarters. Men rushed to their battle stations. They could make out the shape of a large ship moving upriver, just as they had been expecting to see.

Up on the forecastle, Lieutenant Selfridge held the lock string of the 10-inch pivot in his hand. He sighted over the barrel. The gun was laid to fire right into the deck of the approaching ship. No answer had come from the stranger to *Cumberland's* hails. Selfridge shouted, "Shall I fire, sir?"

Commander Rowan, captain of *Pawnee*, heard the question that Selfridge had shouted aft and realized what was about to happen— *Cumberland* was preparing to fire into his crowded decks. At that moment, someone aboard *Pennsylvania,* a receiving ship deemed not worth sinking, hailed them. *Pawnee* heard the hail and replied.

On hearing the ship was *Pawnee*, the men onboard *Pennsylvania* began to cheer, and the cheer was taken up aboard *Cumberland*. The men onboard *Pawnee* returned the cheer with gusto and, one imagines, a bit of relief.

To the men hunkered down onboard *Cumberland* and *Pennsylvania,* *Pawnee* was their salvation. She had come to lift the siege and to tow the *Cumberland* to Fortress Monroe. Pendergrast would later testify that if they had known *Pawnee* was on her way, they most likely would not have scuttled the ships, since the steamer provided a means of getting them to safety.

Pawnee came alongside the shipyard's wharf, and Wilkes was sent to locate McCauley. When he did, he found McCauley was "armed like a Brigand—swords and pistols in his belt and revolvers in his hands." Wilkes led McCauley back to *Pawnee*. The old man was drunk and confused. "His [McCauley's] condition was such as unabled him to comprehend the situation and great difficulty in walking steady. He was unfitted for all duty."

Wilkes and McCauley met Commodore Paulding and several of the other officers on the wharf. Paulding, who had come with the intention of getting all the ships out, was not in the least happy to find they had been scuttled. He relieved McCauley of his command. McCauley staggered back to his private quarters, finished forever with the Gosport Naval Shipyard.

Paulding turned to Pendergrast and asked why the ships had been sunk. Pendergrast repeated their fears of batteries across the river and thousands of men massing to take the yard. Paulding had no reason to doubt this assessment, but he still hoped to get the ships out if at all possible. He dispatched one of his officers, Commander W. M. Walker, whose job had been to take command of *Germantown*, along with the carpenter from *Cumberland,* to determine if any of the ships could be saved.

Lieutenant Henry A. Wise had been placed on special service, sent aboard *Pawnee* and detailed with helping to get *Merrimack* to Hampton Roads. He stepped ashore with Commodore Paulding and with the commodore learned that the ships had been sunk. Hoping it was not too late, Wise raced across the yard, following the river, until, 300 yards away, he came up to the sinking *Merrimack.*

She was low in the water, but she was not submerged. Wise got onboard her and climbed down to the lower deck, trying to determine how far gone she was. He threw a block (one of the big wooden pulleys used in a ship's rigging) into the hold and listened for the splash. The sound that came back to him indicated that the water was over the orlop deck, and the engines and boilers were underwater. It would take more time than they had to get the ship floating again, Wise reported to Commodore Paulding.

At that point, Paulding had been at the shipyard for an hour, and he had decided it was a lost cause. Along with Wise's report, Walker had found that only the little brig *Dolphin* could be prevented from sinking entirely. "I made up my mind that a necessity then existed for destroying the public property to save it from falling into the possession of the people then assembling in great number to take the yard."

Paulding handed out assignments. Lieutenant Wise was to see about burning the ships. Captain Wilkes, who had been sent to take command of *Merrimack*, was to oversee burning the buildings. Captain Wright of the army and Commander John Rogers, who had come to take command of the *Plymouth,* were to blow up the dry dock.

Before the fires were lit, Paulding wanted to destroy all the guns scattered around the yard, realizing that the wealth of ordnance would be as valuable to the Confederates as the yard itself. Pendergrast sent a hundred men from *Cumberland* and they went after the Dahlgrens, including the guns from *Merrimack*'s battery, with 18-pound sledge hammers.

The only way to permanently disable a gun, short of smashing the barrel, is to break off the trunnions, the short arms that project from the side of the barrel on which the barrel rests in the carriage. For an hour the Cumberlands hammered away at the trunnions. The shipyard rang with the sound of their blows, but they managed to break not a single one. The new iron was just too tough. At last they moved on to the gun park and attacked the older 32-pounders and the other old pattern guns and broke those trunnions off "like glass." The guns that could not be destroyed were spiked—a metal spike was pounded into the touchhole, where the fuse to ignite the charge was placed. But that was only temporary damage, easily repaired. Paulding would leave a fortune in guns for the Southerners.

Finished with the guns, Lieutenant Wise set about preparing the ships for the torch. He went aboard *Merrimack* and for some reason—sentimentality, souvenir hunting, the need to save something—he took a binnacle lantern out of her. He then reported to *Pawnee*, where he was given a boat loaded with the turpentine, cotton waste, and other in-

flammables they had received from *Anacostia*. Paulding gave Wise a list
of the ships to be burned, and the lieutenant was off.

On board *Merrimack*, with the help of another officer and some of
the boat's crew, Wise collected up whatever would burn, cordage, rope,
ladders, gratings, and laid it all out on the gun deck in the form of a big
V from the mainmast to the gun ports on the river side of the ship. On
top of that he laid cotton waste saturated in turpentine, with the end of
the cotton hanging out of one of the gun ports. He splashed more turpentine on the decks and beams.

Wise and his crew finished up with *Merrimack* and then prepared
the other ships in similar manner. Even as they worked, they could see
more and more locals gathering on the shore, and boats moving in the
dark. The threat was imminent. The Union men moved quickly.

By one-thirty in the morning, all the vessels worth burning were
prepared for the torch and Wise returned to *Pawnee* to report. Paulding
ordered Wise to wait for a signal rocket before setting the ships ablaze.
Both *Pawnee* and *Cumberland* were near the ship houses and downwind of them. Paulding intended to get the ships well away from the
shipyard before it erupted in flames.

Wise returned to his boat. It was a clear, starlit night. On the far
shore, in the town of Norfolk, the lieutenant could see mobs of people
gathering by the river, and boats were keeping close watch upstream.
Wise kept his boat in the shadows, moving from the shelter of one sinking vessel to the next, trying to present a difficult target to the men onshore, waiting for *Pawnee* to fire her signal.

Beyond the Shipyard Walls

It was clear to General Taliaferro of the Virginia Provisional Army what
was going on in the shipyard, and what would happen next. No doubt
he deduced that if the Yankees were scuttling their ships and destroying their guns, they believed themselves faced with immediate and
overwhelming attack.

Playing on that fear, Taliaferro sent a delegation to the shipyard under flag of truce and headed up by the mayor of Norfolk. They offered Paulding a deal from Taliaferro: to "save the effusion of blood," the Virginians would allow *Cumberland* (and presumably *Pawnee*) to leave unmolested if the Union men did not destroy the shipyard.

It was a clever and daring ploy, but Paulding was not so panicked as to fall for it. Paulding understood that with *Cumberland* and *Pawnee*'s firepower, he was still in a very strong position, no matter how many Virginian troops were gathered at the gate. He told the deputation that "any act of violence on their part would devolve upon them the consequences." The message was clear. The towns of Norfolk and Portsmouth were at the mercy of the navy's guns.

Destruction of the Yard

As valuable as anything else at the Gosport Naval shipyard was the great granite dry dock, and Paulding was eager for the secessionists not to have it. Along with Captain Wright of the army engineers and Commander John Rogers, Paulding sent forty of the Massachusetts infantry and a few sailors from *Pawnee* to blow up the huge gate at the water end, which would render the dry dock useless.

Fortunately for Wright and Rogers, they found the dry dock empty, with only a couple of feet of water at the bottom. Near the river end of the dry dock was a pumping gallery that formed a convenient tight place for exploding gunpowder with maximum destructive force. There, using whatever material they could lay their hands on, the men built a platform to keep the gunpowder above the level of the water. On top of that platform, and in the pumping gallery, they stacked two thousand pounds of gunpowder, enough to destroy the dry dock gate and possibly a good deal more.

With the powder in place, Wright led a train of gunpowder from the gallery to a distance calculated to give them time to escape. (General Taliaferro would later claim it was timed to explode after the Virginians had entered the yard, but that does not seem to have been the case). Into the train of powder they thrust four lengths of slow match.

With preparations complete, Wright sent all the men back to the ships, save for one seaman from *Pawnee* who was kept as lookout for the signal rocket. Then, like Lieutenant Wise, floating in the dark, Wright and Rogers waited.

Charles Wilkes was also waiting. Wilkes had been charged with overseeing the destruction of the shiphouses, marine barracks, storehouses, shops, and other buildings. He in turn ordered Alden and Commander B. F. Sands to prepare the buildings for burning. Around 1:45 A.M., on the morning of April 21, Alden and Sands reported that the buildings were ready to go.

All the men in the shipyard—the Massachusetts volunteers who were spread out in a defensive position, the sailors, and the marines—were all ordered back to the ships. They made their way across a yard lit by the flames of the burning marine barracks, which had accidentally been set on fire and allowed to burn. Left behind were those men whose job it was to actually ignite the various incendiaries.

As *Pawnee* was preparing to cast off, Commodore McCauley's youngest son appeared on deck, tears running down his face. Desperate with fear for his father's life, the boy told Paulding that the old commodore refused to leave his quarters, even after his son had implored him to go. The old man, it seemed, was determined to go down with his command.

Paulding sent Alden to bring McCauley to the ship. Alden found the commodore lying on a couch in his quarters, stupefied by drink and by events moving too fast for him. McCauley was "overcome with chagrin and mortification" that the yard was to be burned with no effort made toward defense. Alden told the commodore it was time to go, and this time McCauley listened. Alden helped him back onboard *Cumberland*.

At around 2:25, the crew of *Pawnee* began winding the ship. This proved more difficult than expected, and it was another hour before she was turned and ready to steam out into the river. Another forty-five minutes were consumed in passing hawsers to *Cumberland* in order to tow her away. Finally, at 4:15, with towlines from *Pawnee* aboard and the tug *Yankee* alongside, the *Cumberland* slipped her mooring cable and was under way.

At 4:20 the rocket was fired. Minutes later, the Gosport Naval Shipyard was turned into a scene from hell.

Chapter 6

The Fall of Norfolk

Lieutenant Wise saw the rocket lifting up from *Pawnee*'s deck. He touched off the combustibles onboard *Merrimack* first. Doused with turpentine, the ship ignited "like a flash of gunpowder."

Just seconds after Wise lit the cotton waste he had left hanging overboard, flames burst from every part of the ship, gushing out the empty gun ports, lifting up through the hatches to the spar deck, racing along the ship's 275-foot length. Wise ordered the men at the oars to pull and pull hard. They barely escaped the destruction that they themselves had brought about.

Wise moved on to the other ships, *Germantown, Raritan, Columbia*, and the brig *Dolphin*, setting them off one after another. Coming alongside *Plymouth*, he found that that ship had sunk so far that the powder train was underwater. There was no time to lay another one. Every second Wise expected to hear the dry dock explode, expected great blocks of granite to come raining down on him and his boat.

At last they crossed the channel and torched *Pennsylvania. Pennsylvania* was a ship of the line, once the mainstay of war at sea. But the days of ships of the line were already over, a fact that this new war would prove decisively. With *Pennsylvania* burning well, Wise turned the cutter downriver and struck off for *Pawnee*, which was already lost from sight.

Shiphouses

Commanders Alden and Sands also saw the rocket, and they set their torches to the buildings they had prepared.

Like the ships, the buildings ignited with frightening speed. Just minutes after they were set on fire, the storehouses, ships, and massive A-frame shiphouses were lost in a sheet of flame, rising 60 feet and more above the shipyard. The choking smoke swirled around, making it hard to see. Alden and Sands and the men with them hurried across the burning yard, reaching the boat that was to take them to safety just as the entire yard was lost in a wall of flame.

In command of that boat, moored at the dock near the ship houses, was the much-disgusted Captain Charles Wilkes. It was his job to collect the eight men still ashore and bring them back to *Pawnee* once the shipyard was fully involved.

Wilkes had positioned his boat just ahead of the now-burning *Germantown*. Astern of *Germantown* was *Merrimack*, lost in a tower of flames. Alden and Sands and the others reached the dock and took their places in the boat just as the fire was climbing over the ships' rigging and spars, the flames reaching even higher in the air, the smoke rolling over them. The shipyard was brilliantly illuminated by the blaze. The men in the boat looked anxiously toward the dry dock, waiting for Wright and Rogers to get there. They did not want to leave the men behind, but they could see nothing beyond the flames.

Prisoners of War

Wright and Rogers also saw the rocket rising up against the backdrop of stars. On seeing the signal, the two men touched off the four lengths of slow match thrust into the powder train and then climbed

up out of the dry dock. They made their way across the yard, heading
for the place where Wilkes had arranged to meet them.

All around them, buildings were bursting into flame, and along the
seawall and out on the water the ships were engulfed. As they hurried to
the rendezvous they were met with flames spreading across their path.
Soon there was nothing but fire in front of them, a wall of flame, and the
heat was intense, so intense that they did not believe the boat could still
be there, that it "had undoubtedly been driven off." And even if it was still
waiting, they had no way of getting to it. They were cut off from the river.

The officers abandoned hope of reaching the boat and began looking
for some other way out of the burning shipyard. After stumbling around
for some time, they finally came to the western gate in the brick wall surrounding the yard. The wooden gate, like the rest of the yard, was on fire,
but they managed to get through it to the relative safety of Portsmouth.

Making their way to the riverfront, Wright and Rogers found a boat.
They climbed aboard and took up the oars, hoping to pull unnoticed
down river to the Union ships. But there was no cover of darkness now.
Before they had gone very far, they were spotted in the brilliant light of
the burning yard. Small-arms fire crackled from the shore, and musket
balls slapped the water around them. Worse, they could see an armed
mob forming at a point downstream of them where the river narrowed
to within easy musket shot. Knowing they could not run that gauntlet,
they crossed the river to Norfolk and delivered themselves into the
hands of General Taliaferro, prisoners of war.

Union Retreat

By the time Wright and Rogers had raced through the burning gate,
Wilkes and his boat were long gone. They had waited as long as they
conceivably could, waited while the *Germantown* and the *Merrimack*
were consumed by fire. They waited until the flames of the shipyard
were so intense that their faces were scorched with the heat and they
could no longer bring the boat alongside the dock.

Blinded by the flames and the thick smoke that blanketed the yard,
Wilkes ordered the boat under way. As they pulled away from the smoke

they met Lieutenant Wise's cutter, which seemed to emerge right out of the wall of flame. Together, with "large flakes of fire falling around us," the boats headed down stream.

Wilkes had believed that *Pawnee* and *Cumberland* intended to cross over to Norfolk and anchor, but they did not. Instead, the ships continued on downriver, making for Hampton Roads. The two boats had a long pull to catch up with the ships, but they did at last, near Craney Island and the mouth of the Elizabeth River. The way out of Norfolk was well lit, at least until the burning yard was well astern, and the night echoed with a series of explosions as the fire found shells and gunpowder. Fortunately for Wilkes and Wise, the militia who would later force Wright and Rogers to surrender had not yet assembled.

With the sun coming up, the men onboard the Union ships could see the great cloud of black smoke rising up from the navy yard about six miles upriver from them. Ahead they found that the Virginians had sunk even more obstacles in the river. Paulding did not think he could get the deeper *Cumberland* over them. He anchored *Cumberland* upstream of the obstacles and with *Pawnee* steamed on alone to Fortress Monroe, where he discharged the Massachusetts volunteers.

Waiting at Fortress Monroe was the steam tug *Keystone State*, which had been hurriedly fitted out and loaded with over one hundred sailors and marines to help with the evacuation of the navy yard. It was too late for that duty, but Paulding sent the tug to assist in getting *Cumberland* over the bar. It took several hours of pulling, but in the end *Keystone State* and *Yankee* together were able to haul the sailing man-of-war over the underwater obstructions and tow her to the safety of Fortress Monroe's guns.

Paulding transferred his flag to *Keystone State* and steamed back to Washington. Arriving at the Washington Naval Shipyard, he discovered that all but two of the officers there had resigned and gone South, the mail interrupted, the telegraph wires cut, and the lightboat on the Potomac River burned. The nation was at war, and the most important shipyard in the Union was lost to the enemy.

The Enemy at the Gate

The question remains: how real was the threat to the Gosport Naval Shipyard? Did Paulding get out in the nick of time, saving everything that could be saved, as he contended? Or was it a panicked debacle, a "great want of common sense," a gift to the Confederacy, as Captain Charles Wilkes would later suggest?

From the onset of the secessionist debate in Virginia, those who advocated parting with the Union understood that there were two United States facilities that they had to get their hands on: the federal armory at Harper's Ferry and the Gosport Naval Shipyard.

On April 17, while a Richmond convention was still debating whether Virginia should secede, ex-governor Henry Wise announced that militia groups were at that very moment seizing Harper's Ferry and the shipyard. He was wrong, but not so very wrong. Wise himself had met the day before with officers of the state militia, ordered them to take the facilities, and sent them on their way. He did not wait for permission from then-governor John Letcher, whom he considered not staunch enough a secessionist. Letcher agreed to the deployment after the fact.

Major General William B. Taliaferro was dispatched to Norfolk on the night of the 18, arriving there the following morning. With him went his aides, Captain Henry Heth and Major Nat Tyler, as well as two naval captains freshly appointed to the Virginia State Navy, Robert Pegram and Catesby ap Roger Jones*. The former was sent with orders to assume command of the naval yard once it was in Virginian hands.

They found the city in an uproar. Weeks of smoldering excitement had been ignited by the sinking of the lightboats in the channel, the first concrete action taken to prevent the ships at the yard from leaving. They found that the militia had been turned out. They found that the people of Norfolk had turned their city's governance over to a committee of safety, better able to handle the military crisis.

They did not find much else.

---

*The "ap" in Jones's name is a Welsh convention meaning "son of," in this case, Catsby, son of Roger Jones.

The militias of Norfolk and Portsmouth, the Blues and the Grays, respectively, numbered only around 600 men, and few of them had weapons. The only artillery consisted of a few 6-pounder fieldpieces. There was nothing that could be considered a naval force.

Taliaferro estimated that the number of men defending the yard was about the same as the militia under his command. But the defenders were well armed, and the navy yard had ships, their batteries lining the river, able to fire either into the shipyard, should that be attacked, or across the river at the town of Norfolk. The flat local geography (many people called it "champagne country") meant that the big naval guns had a clear shot in any direction.

Taliaferro knew that taking the yard would be no great feat—the wall surrounding it presented a minor obstacle, and the marines defending it were few. However, once inside the yard, the militia would be cut down like deer in a pen by the ship's broadsides.

"To have boarded the ships," Taliaferro reported, "would have required, by the estimate of some of the most experienced naval officers, at least 800 men—sailors, or such men as were familiar with boats." Captain Pegram felt they should not attack the yard and ships with less than five thousand men. Other officers felt more were needed.

Despite the "excitement and eagerness of some ill-advised persons for immediate hostile action," Taliaferro did not think the time was right to attack. Though McCauley and Pendergrast were certain he was setting up batteries to fire on the yard, Taliaferro was in fact concerned about his lack of ordnance. He telegraphed Governor Letcher and requested that heavy guns be sent to the area so that he could establish the very batteries the Union officers thought he already had.

Taliaferro also ordered reinforcements from Petersburg, a battalion that did not arrive until the night of the 20th. Rather than use the new troops in the offensive, Taliaferro was more concerned about his back door. He stationed the Petersburg men in the rear of the city of Norfolk, to defend against federal soldiers from Fortress Monroe who could be landed near Sewell's Point and marched overland.

Commodore McCauley would later report that "two thousand men came down on Friday, April 19, from Richmond, Petersburg, and other places in the vicinity of Norfolk," and that "there was no doubt of the fact that they were erecting batteries under cover of the heavy pine woods . . ." Little wonder that he felt compelled to scuttle the ships in the yard. But where had he received his information?

All of the intelligence upon which McCauley and Pendergrast relied was provided to them by the citizens of Portsmouth and Norfolk. McCauley employed a number of local men as messengers to bring mail from Portsmouth and to communicate with the town. The messengers also brought wild stories about troop buildups, militia pouring in from all over the state, and, yes, batteries being erected in piney woods. Some of those locals coming and going from the yard were also stealing small arms.

It did not occur to McCauley and Pendergrast that the Virginians could be purposefully feeding them disinformation. In fairness to the Union officers, they had almost no other way beyond taking the word of locals to discover what was going on beyond the walls of the shipyard. At one point McCauley did send Lieutenant Selfridge to Norfolk to try and see for himself what was going on, but Selfridge was not able to discover anything, possibly because there was nothing to discover. Also, the presence of Major General Taliaferro seemed to have impressed the officers and helped them believe there were large numbers of troops massing at the gate.

The sound of trains running in and out of Norfolk and Portsmouth, whistles blowing and the shouts of disembarking men, also helped convince the Union men that the enemy was massing for an assault. It seemed to have quite unnerved McCauley but it was, according to Confederate naval officer turned historian J. Thomas Sharf, a *"ruse de guerre."* The president of the Norfolk and Petersburg Railroad ordered his trains to run empty for a short ways out of the city and then return with a few local citizens aboard who whooped it up like soldiers on their way to a fight. He kept these theatrics up all through the day and evening of Friday the 19th, and apparently even after that. Lieutenant Wise reported hearing trains on three occasions as he was waiting to set *Merrimack* on fire.

All told, it was an amazingly effective campaign of disinformation, and it had the desired effect. Even when Taliaferro told Selfridge, truthfully, that there were no batteries, and Heth reiterated it to McCauley, the Union officers did not believe it. They were convinced that an overwhelming force was about to be unleashed upon them, and the best they could do was to destroy the yard and run.

Paulding, showing up well after the Virginia troops were supposed to have arrived, could only take McCauley and Pendergrast's word. In any event, he was not left with much choice regarding the ships. Wilkes had advocated at least waiting until daylight to get a better sense for the threat arrayed against them. He pointed out that the only "troops" they saw were "not even half a dozen men who were apparently lounging about the Gate from curiosity." In retrospect, his argument seems to have been the correct one.

By Taliaferro's own estimate, the Virginians were in no position to attack. His offer to Paulding on the night of the 20th, to allow *Cumberland* to leave unharmed if they did not burn the shipyard, was pure bluff. He had no means of injuring *Cumberland*, while *Cumberland* posed a serious threat to him. The U.S. forces in the yard certainly could have held off attack for a day or two while they loaded the more valuable tools and ordnance on shipboard and towed it away. Selfridge would later say, "I do not hesitate to state that the Yard could have been held by the *Cumberland* and *Pawnee* . . . until the ordnance stores and the *Merrimac* could have been removed or effectively destroyed . . ." The whole thing was, in his mind, "an irretrievable blunder." Others would have harsher assessments.

With the shipyard abandoned, Taliaferro ordered the Portsmouth militia to take possession and make an effort to extinguish the flames. Early in the morning of Sunday, April 21, the Virginians entered the yard and claimed it for themselves.

What they found was a bounty of war matériel. Taliaferro reported, "The damage was not so great as that at first apprehended. Only an inconsiderable portion of the property, with the exception of the ships, was destroyed, and some of the ships may yet be made serviceable."

The dry dock had not blown up. It is not known for certain why the gunpowder failed to explode. One report has it that the presence of the explosives was reported to Lieutenant C. F. M. Spottswood, who promptly opened the gates and flooded the dry dock, submerging the mine before it could detonate. Spottswood was one of the naval officers stationed at Gosport who chose to resign and go South, so it is reasonable to assume he knew how to operate the dry dock gates. However, it seems likely that if the powder train had remained lit, the mine would have gone off before Spottswood was able to flood it. Most likely the powder train failed to ignite for some other cause.

By U.S. government estimates, the value of everything left to the Virginians—and soon after the Confederate Government—carried a total value of nearly five million dollars.

The U.S. navy left for the Southerners 130 gun carriages and over a thousand guns, from 11-inch to 32-pounders. They left most of the machinery in repairable condition. They left two thousand barrels of gunpowder, thousands of cartridges, thousands and thousands of shot and shells.

And they left the *Merrimack*.

Chapter 7

The Confederate States Navy

The Confederate States Navy was just two months old on the day that it inherited the most extensive naval yard in America.

On February 20, "An act to establish the Navy Department" was passed by the government of the Confederate States of America, which was itself created earlier that month. The brief act called for a chief officer to be known as "Secretary of the Navy" who would "have charge of all matters and things connected with the Navy of the Confederacy," under the control and direction of the president. It also authorized the secretary to appoint a chief clerk and other clerks as needed.

The salary for the secretary was later set at $6,000 per year. The total operating expenses for the navy for the first year were estimated at $2,065,110. As far as the allocation of funds, only the executive mansion was given less money than the navy.

The secretary of the navy had a minuscule staff. There was no assistant secretary of the navy, as Gideon Welles enjoyed, no Board of Admiralty as in the British navy. A disproportionate amount of administrative work and decisionmaking would fall to one man. Jefferson Davis needed the right man.

On February 25, Davis informed the Confederate Congress that he had appointed the former U.S. senator from Florida Stephen R. Mallory to the position. This announcement did not meet with overwhelming enthusiasm.

Fairly or not, Mallory was viewed by many as being lukewarm to secession. True, he was not a "fire-eater," one of those who had been calling for the secession of the southern states years before Lincoln's election. It was always his desire to see southern concerns—for which he was an unwavering champion—rectified through means less radical than opting out of the United States.

It was not until Florida's secession was all but a *fait accompli* that Mallory sided with those who wished to leave the Union. In January 1861 he joined a steering committee of U.S. senators such as Jefferson Davis and John Slidell that passed a resolution calling for the secession of the Southern states.

Mallory was also considered suspect for his part in handling the standoff at Fort Pickens, a federal fort that overlooked the Pensacola Navy Yard in his native state of Florida.

On January 12, 1861, rebel militia captured, without a fight, the small navy yard from the federal troops stationed there. Like Anderson at Sumter, Captain Adam Slemmer, commanding the Union troops at the navy yard, retreated to Fort Pickens, situated on Santa Rosa Island about a mile offshore. The standoff began.

The yard itself was fairly useless if the enemy controlled the guns of Fort Pickens that commanded it. Initially, Mallory had been in favor of capturing the fort along with the navy yard. But with federal troops occupying it, it was a different matter. Capturing Pickens would be a bloody business, and neither North nor South wished to draw first blood. A number of Southern senators, Mallory among them, wrote to the commander of military forces in Florida instructing him not to attack Fort Pickens. "Blood shed now may be fatal to our cause," they warned.

Mallory, along with Captain Samuel Barron of the U.S. Navy (and soon after the Confederate States Navy) negotiated a truce with President Buchanan and his secretary of the navy, Isaac Toucey. If the federals did not reinforce Fort Pickens, and the U.S. naval vessels did not try to enter Pensacola Harbor, then the rebel forces would not attack. It was a good solution given the circumstances, but many in the South saw it as cowardice and betrayal and they were not ready to forgive Mallory.

For all the trouble Mallory had at the onset of his appointment, he was an ideal choice. Insightful, creative, an excellent administrator, and a man of vision, the Confederate navy could have done no better than it did under his able hand. Somewhat round and stocky, with unruly hair and a beard that lined his chin like an Amish farmer's, Mallory was neither colorful nor flamboyant, but then the Confederacy had all the flamboyance it needed. Mallory went quietly about his job, and he did it well.

The Making of a Senator

Stephen Russell Mallory was born on the island of Trinidad in 1811, the second of two sons. His mother, Ellen Russell, Irish by birth, was sent to Trinidad as a girl to live with her uncles, who were planters on the island. There she married a Connecticut construction engineer, John Mallory. Ellen was Catholic, and she would raise Stephen in that faith. Mallory would be the only Catholic in the Confederate war cabinet.

Soon after Stephen's birth, the family moved, first to New York, then to Mobile, Alabama, staying only briefly at each place before settling in Key West, Florida. At nine, Mallory was sent to a school near Mobile, which he attended for about a year. He learned to read there, but not much else, spending the bulk of his time in the woods or on the beach, riding, swimming, and shooting.

Mallory's time at school ended with the deaths of his father and older brother. Stephen returned to Key West to help his mother, who had opened a boardinghouse to support herself and her remaining child. Soon after, Mallory was again sent away to school, this time to the academy of the Moravians in Pennsylvania. The Moravians were a bit stricter in their pedagogy than Mallory's former school, and there he received a grounding in Latin, Greek, music, and other disciplines. Mallory stayed at the academy for three years, until his mother could no longer afford to keep him there. It was all the formal education he would ever have.

Mallory returned to Key West. He enjoyed the outdoor life, hunting and fishing, and learned woodcraft from local Indians with whom he

was acquainted. In 1830 he became the inspector of customs, a position that sounded more important than it was. He also undertook at
that time to complete his education on his own, and began reading
widely. He developed a career strategy—"study law, become a lawyer,
and at some day go to Congress."

By 1836, Mallory was starting to make a name for himself in local
politics. That year he wrote to Angela Moreno, a friend of his cousin's,
renewing a proposal of marriage he had made two years before, on the
occasion of their first formal meeting (Mallory's first proposal had been
more of an announcement of intent). This time Mallory informed Angela that if she did not marry him, he would join the army and help fight
the Seminole Indians in South Florida. Angela replied that that was a
wonderful idea.

For the next two years, Mallory commanded a schooner-rigged cen-
terboard whaleboat, which he named the *Angela*. With a handful of
sailors as crew, he cruised the coast in search of the enemy. In reality,
Mallory's service had all the markings of a two-year camping trip in the
wilderness he so loved.

Mallory's years of steadfast devotion eventually worked away at Angela's resistance. They were married in 1838 and continued on in Key
West. By then Mallory had his own legal practice. Like most lawyers in
Key West (and there were quite a few), his work centered mainly on
maritime law, including disputes between shipowners and salvagers. It
was his first real acquaintance with the maritime issues that would become central to his life.

Mallory continued to rise in prominence in the Democratic Party,
gaining a reputation for intelligence and good sense. In 1851 he was
nominated and elected by the state legislator for the U.S. Senate before
he even knew he was a candidate.

Mallory's election devolved into a contentious fight with fellow
Democrat David Yulee, ending in a disputed vote. In one of the first of
what would become a tradition of botched elections in the state of
Florida, the U.S. Senate itself actually decided in the end which of the
candidates would take the seat. It was Mallory.

During Mallory's tenure as a senator, he served alongside some of
the most notable politicians in American history. There was Stephen

Douglas, Charles Sumner, Robert Toombs of Georgia, and, from Mississippi, Jefferson Davis. The wily William Seward, the future secretary of state for Abraham Lincoln and arguably the most powerful man in that administration after Lincoln himself, was also part of Mallory's first Senate class.

This impressive body of men met in an atmosphere of extreme tension. 1850 had seen long and acrimonious debate over a bill aimed at cooling the superheated cries for secession. The Compromise of 1850 ultimately passed, though it was hardly a compromise, nor did it do much beyond put off for a decade the inevitable split between North and South.

Mallory's maiden speech concerned naval matters, when he rose in favor of reinstating flogging as a means of keeping discipline onboard men-of-war. Flogging was not reinstated, but Mallory continued to be active in naval affairs. In 1853 he was appointed chairman of the Naval Affairs Committee of the Senate, a committee on which he already sat. He approached his position with fervor and enthusiasm.

That same year he helped Secretary of the Navy Dobbin push through a bill to revitalize that moribund service, including the building of six first-class screw frigates and six steam sloops of war. The larger vessels, which included *Minnesota, Colorado,* and *Roanoke,* were launched in 1855 and 1856, sailing frigates with steam auxiliary. They were the finest examples of such ships in any navy in the world. When the first of them traveled to Europe, she was much admired. The British navy extended that sincerest form of flattery and imitated her design in several of their own subsequent vessels.

Eight years later, these would be the very ships against which he would have to fight, with the exception of one of the frigates, the U.S.S. *Merrimack.*

Mallory also championed the completion of the first ironclad floating battery, the "Stevens Battery," begun as a naval experiment in 1842. Still not completed a decade later, the Stevens Battery had run through its funding, and despite Mallory's advocacy, the Senate was not willing to pour more money into it. Still, Mallory's position was indicative of his forward thinking in regard to naval technology, an open-mindedness that would be crucial in the resource-starved Confederacy.

Mallory's most controversial work was as sponsor of the Naval Retiring Board, an effort to prune the considerable deadwood out of the navy. The board, which met in secret, was comprised of several men who would play a big part in the coming war, including McCauley, Pendergrast, Samuel Du Pont, Samuel Barron, and James Buchanan.

Though the board was generally successful, it came under terrific criticism and was forced to overturn many of its decisions. In the course of the fight, Mallory made more than a few enemies, including the brilliant oceanographer Matthew Fontaine Maury, whom the board attempted to retire. Some of the officers selected for pruning, including Maury, would later become part of the Confederate Navy, bringing their enmity with them when they went South.

In 1857 Mallory was elected for a second term in the Senate, which he did not complete. On January 10, 1861, Florida left the United States and Mallory went with his state. Even the methodical, organized, carefully planning Stephen Mallory, who as a young man had mapped out the trajectory of his career, could not have envisioned the turn his life was about to take.

Chapter 8

A Chieftain Without a Clan

Stephen Mallory was better prepared to assume the role of secretary of the navy than was his Union counterpart, Gideon Welles. After his years as the Senate's leader on naval issues, Mallory had already established a personal philosophy on the use of naval power. Ever an autodidact, he had educated himself on naval affairs with the same zeal that he had taught himself the law and, indeed, nearly everything else he knew. He was familiar with the state of the art of naval warfare. He understood at the onset of the war, better than Welles did, the role his navy would play in the conflict.

Mallory may have had a philosophy, but he did not have ships. In that he was compared by a contemporary to "a chieftain without a clan, or an artisan without the tools of his art." The Confederate navy at its inception consisted of a few small vessels turned over by the states and those captured at Pensacola. To get a real navy, Mallory would have to build one or buy it. His plan was to do both.

As early as the middle of March, Mallory sent naval officers to Southern ports as well to ports in the United States and Canada in search of merchant vessels that could be purchased and converted into men-of-war, but they met with little success. Nor were converted merchant ships the kind of navy Mallory had in mind. His ambitions ran to loftier heights.

While Mallory was still in the U.S. Senate, ironclad technology had begun its ascendancy. This was particularly true in the nations that wielded real-world power—England and France—where an arms race over ironclad ships was under way.

The State of the Ironclad Art

The first impetus to clad ships in iron plating came in the 1820s with the development of the shell gun for naval use. Wooden shipbuilding had advanced to the point where line-of-battle ships were reasonably well protected against round shot, even fired from a few yards away. But they were no match for exploding ordnance, with its vastly greater range and destructive power. When exploding shells were combined with the accuracy of rifled guns, wooden ships were helpless. The great leap forward in artillery technology required a similar leap in the ability of warships to resist it.

That leap began in the United States in 1842 with the development of the Stevens Iron Battery, which was never completed, despite Mallory's advocacy of the project. Nonetheless, tests done on the battery proved the potential invulnerability of iron plating over a thick backing of wood. Though the U.S. government paid little attention, the governments of England and France did, and began experiments of their own, the French in particular.

The ability of iron plating to resist shot had its first real-world test during the Crimean War. But before the need for such technology was clearly understood, a few military disasters were required.

In 1853, a Russian fleet of nine ships engaged a Turkish fleet of thirteen. The Russians were armed in part with shell guns. They destroyed all but one of the Turkish vessels.

The French and British did not learn from the Turks' mistake. The following year, in alliance against the Russians, ships from the two countries attacked a fixed fortification in Sevastopol with a squadron of wooden ships. The Russian fort was armed with shell guns and they dished out heavy punishment, while the fortress suffered none in return. Wooden walls simply could not stand up to exploding ordnance.

For a year prior to the battle at Sevastopol, the French had been developing floating ironclad batteries. In 1855 they brought them into action against the Russian fortifications at Kinburn at the Dnieper River entrance to the Black Sea. Three of the French batteries were anchored a mere 800 yards from the fortification. At 9:30 the French opened up and the Russians replied in kind. For four hours the two sides kept up a furious exchange.

Accuracy was not a problem at so short a range. Shell after shell slammed into the ironclads and bounced harmlessly away. When the firing was done, the French floating batteries were undamaged, save for a few dents. The Russian fortifications were battered to pieces. It was another step forward in the arms race.

Both the French and the British were looking to iron as part of the next generation of warship. In fact, by the mid-nineteenth century, much of Europe was engaged in a frantic arms race of heavy ordnance versus iron-cased ships. But the lumbering, unseaworthy floating batteries were not the thing to bring naval power to bear around the globe. What was needed were seagoing vessels employing sail and steam and incorporating ironclad technology.

The French were the first to build such a ship with the 255-foot, 5,630-ton, three-masted screw steamer *La Gloire*. *La Gloire*'s hull was wood, covered with iron plating 4.7 inches thick. She mounted thirty-six 6.4-inch guns. For thirteen months she was the most formidable warship in the world.

The following year the British launched the monstrous HMS *Warrior*. At 418 feet in length and 9,137 tons displacement, she dwarfed *La Gloire* and assumed the status of the world's most powerful ship. *Warrior* was of all-iron construction, with the central part of her hull, which housed the main guns, built as a discreet box, or citadel, of 4½-inch-thick iron. For both the French and the British, these ships were only the start of an ambitious program of ironclad ship building.

An Ironclad for the Confederacy

Stephen Mallory was perfectly aware of all this before taking office, and he made an effort to make others, most notably Jefferson Davis and Charles M. Conrad, chairman of the Committee on Naval Affairs, aware of it as well. He also knew, from his Senate days, that the United States Navy lagged far behind in embracing this technology. While Europe was advancing the ironclad revolution, the United States navy had largely ignored it, save for the forgotten Stevens Iron Battery. Mallory saw in that failure a potential edge for the Confederate navy.

Upon assuming office, Mallory moved with speed and conviction, laying out his long-range plans in early April, while the Confederate government was still located in Montgomery, Alabama.

One of the "several classes of ships" Mallory envisioned consisted of fairly conventional, moderate-sized vessels of sail and steam, propeller-driven, to be used as commerce raiders. In early May, Mallory dispatched the highly competent Commander James Bulloch to England to buy or have built six such vessels. "Large ships are unnecessary for this service," he told Bulloch, "our policy demands that they shall be no larger than may be sufficient to combine the requisite speed and power, a battery of one or two heavy pivot guns and two or more broadside guns, being sufficient against commerce." Mallory preferred six moderate-sized ships to a few large ones. Among the ships that Bulloch was able to acquire were three that would achieve lasting fame—the *Shenandoah*, *Florida*, and *Alabama*.

But commerce raiding was only one part of Mallory's plan. Another was raising Lincoln's blockade, which Mallory knew would be a serious threat to the Confederacy. He wrote, somewhat hyperbolically, "I propose to adopt a class of vessels hitherto unknown to naval services." He was thinking about ironclads.

Writing to the Committee on Naval Affairs, he stated, "I regard the possession of an iron-armored ship as a matter of the first necessity. Such a vessel at this time could traverse the entire coast of the United States, prevent all blockades, and encounter, with a fair prospect of success, their entire Navy." The Union navy, Mallory knew, had no ship that could take on an ironclad vessel.

In this instance, however, Mallory was not thinking of an ironclad such as CSS *Virginia*, but rather something along the lines of *La Gloire*. He instructed Confederate navy lieutenant commander James North to join Bulloch in London and arrange for the purchase or construction of two such vessels. He included with his orders a description of *La Gloire* and told North, "The Confederate States require a few ships of this description." North was instructed to consult with Captain Cowper Coles of the Royal Navy, one of the early innovators in ironclad design and a rival of John Ericsson's for the claim of having developed the revolving turret. "The *Warrior*, now being built upon the Thames," Mallory told North, "must have your attention." Two million dollars was allocated for the two Confederate ironclads.

Mallory believed, as did many Southerners, that England and France would be disposed to help their cause, if for no other reason than to maintain the flow of cotton from Southern plantations to European mills. But the English and French proved more hesitant than he had expected. The wooden commerce raiders were one thing—they could be built under the guise of being merchant vessels, allowing the European nations to maintain the appearance of neutrality—but ironclads could not be mistaken for commercial shipping. Bulloch and North met with tremendous difficulty in their effort to acquire ironclad cruisers. Ultimately only one such ship reached Confederate hands, the *Stonewall*, which arrived in the Caribbean soon after the war had ended.

At the same time that Mallory was trying to acquire a *La Gloire* of his own, he was also thinking about homegrown ironclads, no doubt closer to the *Virginia* model. Just a few days after ordering North to London, he sent orders to Captain Duncan Ingraham to visit iron rolling mills in Georgia, Kentucky, and Tennessee to see if any of them could roll iron plate 2 or 3 inches thick. He also wanted Ingraham to find the best means for transporting the plate to New Orleans, that city being the most active shipbuilding center in the South.

Mallory could not have been cheered by Ingraham's report. None of the three mills Ingraham visited could roll the plate iron. The best he got was Daniel Hillman and Co. in Kentucky, which thought they might be able to do so in two months. Mallory had earlier determined that plate could not be rolled in New Orleans, either. The secretary might have seen in that a harbinger of things to come.

Mallory was not the only one in the South thinking about ironclads, far from it. Among those considering the new technology was one of the most capable and brilliant officers in the Confederate navy, John Mercer Brooke.

Brooke, described as "taciturn and dreamy," was thirty-seven years old, a native of Tampa, Florida. He joined the U.S. navy at the age of fourteen, and later graduated from the naval academy at the age of twenty-two. Most of his naval service was spent on foreign stations. He was part of the navy's North Pacific Expedition and was with Matthew Perry during the opening of Japan.

Brooke had an inventive and scientific mind. From 1851 until he left the Union navy he was attached to the Naval Observatory, working primarily on charting the Pacific Ocean he helped explore. He developed a deep-sea sounding device for mapping the deep ocean floor, for which the United States Congress, in 1860, awarded him five thousand dollars. When Virginia seceded, Brooke invested the money in Confederate bonds, "as a good Constitutional patriot ought to have done," and never saw it again.

After joining the Confederate navy, much of Brooke's attention was given to the development of ordnance. Though he had had little prior experience in the field, Brooke gave the Confederacy the ubiquitous Brooke rifle, considered one of the finest guns in the world. Brooke's name would become synonymous with guns in the same way that John Dahlgren's would.

Like a number of naval officers, Brooke's loyalties were divided. It was not until April 20, after Virginia left the Union, that he opted to "go south." He submitted his letter of resignation to Welles, but it was not accepted. Rather, his name was "stricken from the rolls of the navy" by direction of the president, an irrevocable censure that Welles had instituted, rather than cordially accepting the resignations of Southern officers as his predecessor had done.

At first Brooke volunteered for the Virginia State navy and was commissioned a lieutenant on April 23. Per order of Captain Samuel Barron, also recently of the U.S. navy, he served as naval aide to General Robert E. Lee. Brooke enjoyed the assignment, but he wanted to be part of the national force. He wrote to Mallory asking for a "commission in the permanent navy of the Southern Confederacy when such navy shall be organized." On May 2 he was made a lieutenant of the Confederate States Navy.

Six days later, about the time that Mallory was preparing his elaborate justification of ironclad ships to the Committee on Naval Affairs, Brooke wrote to the secretary suggesting almost exactly what Mallory had in mind. Brooke pointed out that "an iron plated ship might be purchased in France loaded with arms and brought into port in spite of the wooden blockade."

In early June, the Confederate government moved from Montgomery, Alabama, to Richmond, Virginia. It can and has been argued that there were better choices than moving the government to within easy marching distance of the enemy's main army. But the lure of the Old Dominion, home of Washington and Jefferson, proved too much.

On June 3, Mallory arrived at his new but temporary offices on the second floor of the old United States Custom House in Richmond (soon after, permanent offices were leased at the Mechanic's Institute building, next to Capitol Square). Among his first visitors was John M. Brooke. Brooke described the meeting to his wife: "The Secretary arrived in Richmond from Montgomery on June 3, and I had some conversation with him that night though the hour was late and Mallory was very tired." Though Brooke does not say they discussed ironclads, the chances are good that they did, given their recent correspondence and the thought both men had been giving to the subject.

If Mallory and Brooke did not discuss ironclads that night, they certainly did over the next few weeks. Brooke later testified, "The Secretary and myself had conversed upon the subject of protecting ships with ironclading very frequently." On June 10, Mallory asked Brooke to design an ironclad vessel along the lines they had discussed and to work out the specifications.

Brooke was not a draftsman, but he did come up with rough drawings of his proposed ship. It consisted of a wooden casemate clad in three inches of iron plate, inclined at as steep an angle as could be and still allow the gun crews enough room to work the guns. The casemate's configuration was not particularly original. The guns were arranged in broadsides like a conventional man-of-war, and the idea of inclining the sides had been around for a while. Stevens Iron Battery employed inclined sides, as did the French batteries used in the Crimean War. Even Brooke admitted that there "was nothing novel in the use of inclined iron-plating."

More unique than the casemate was the underwater shape of the vessel. Brooke's idea was to build the hull supporting the casemate with the hydrodynamic shape of a conventional ship, rather than the boxy lines of a floating battery. "It occurred to me," he wrote, "that fineness of line, buoyancy and protection of hull could be obtained by extending the ends of the vessel under water beyond the shield." While earlier ironclad designers had been thinking about floating forts that could at best move awkwardly around a harbor, Brooke was thinking about speed and seaworthiness.

In Brooke's design, the eaves of the iron casemate terminated 2 feet below the waterline, and the sharp bow and stern that extended beyond the casemate were submerged two feet to shield them from enemy shot. To prevent the water from banking up against the casemate while the vessel was under way, which would have slowed her and could have sent water in through the forward gunports, Brooke designed a false bow and stern that would be just above the waterline. These superstructures, made of ship iron, were to be decked over. Since they were not an integral part of the hull, they could be shot up by enemy fire without causing any real harm to the ship.

Brooke showed the plans that he drew—rough drawings of the proposed ship's body, sheer, and deck—to Mallory. Mallory liked what he saw and approved of the design. Months before the U.S. navy would even think about building an ironclad, the Confederates were on their way.

Chapter 9

An Ironclad for the North

When two warring nations have seats of government less than one hundred miles apart, common borders thousands of miles long, a free press, and populations who look alike and pretty much talk alike, military secrets can be tricky to keep.

That certainly was true for the conversion of *Merrimack* into the ironclad CSS *Virginia*. Despite Confederate attempts to keep the effort under wraps, Gideon Welles "contrived to get occasional vague intelligence of the work as it progressed." Yet the potential threat posed by that novel weapon seemed to have made little impression on him.

Stephen Mallory understood that innovation and creativity would be his primary hope in the face of an overwhelming conventional force he could not match. Welles, in possession of that overwhelming conventional force, did not feel as hard-pressed to innovate and create. He had other concerns.

On April 19, the day that Paulding was sailing from Washington to Fortress Monroe aboard *Pawnee*, Abraham Lincoln declared a naval blockade of the lower Southern states. The proclamation read in part, ". . . a competent force will be posted so as to prevent entrance and exit of vessels from the ports aforesaid." That was easy enough to put on paper, but putting such a force on the water was another matter. And it was Gideon Welles's problem.

The problem only got worse on the 27th when Lincoln added Virginia and North Carolina to the blockade. It was a mammoth task for any navy, but the U.S. navy was particularly unprepared. The navy was, by Welles's assessment, "feeble, and in no condition for belligerent operations. Most of the vessels in commission were on foreign service; only three or four, and they were of inferior class, were available for active duty."

Welles began deploying what ships he had and buying more as quickly as he could. Naval officers with little experience in dealmaking were sent to purchase ships of any type that could mount a gun, and those officers were royally and systemically ripped off by sharp New England shipowners. Alerted to the scam by financier friends, Welles finally employed businessmen of his own to act as shipbrokers. Soon the navy was paying a more reasonable price for vessels.

But converting merchantmen into navy vessels was a quick fix, a stopgap. A merchant ship, beefed up and armed with a few guns, was a far cry from a built-for-the-purpose man-of-war. So along with buying, the navy department began an energetic building campaign. Within four months of the beginning of hostilities, the Union had forty-seven vessels under construction. But not one of them was an ironclad.

Despite his lack of interest in ironclads, Welles could not ignore them completely. Welles was aware of the revival of *Merrimack* as well as Stephen Mallory's efforts to obtain ironclads from the British or French. It was a threat that had to be matched.

Welles and Gustavus Vasa Fox began to acquaint themselves with the state of the art. Fox was the assistant secretary of the navy (a position created specifically for him), a dynamic, highly competent "live man," as Lincoln called him. The two men read what they could about armored vessels. They consulted with senior navy men, most of whom felt that ironclads were, as John Lenthall, Chief of Construction, Equipment and Repair, called them, a "humbug."

Welles at least had the trait, shared by Mallory, of not being tradition-bound, as senior naval officers tended to be. Despite many of his advisers' dismissal of ironclads, Welles felt his department should investigate the new technology. But he still did not feel the same sense of urgency as his opposite number in the Confederacy. Congress was to reconvene on July 4, called by Lincoln into special session to deal with a crisis quickly blooming into civil war. Welles let the issue wait until then.

It was not until the second or third week of July, more than a month after the Confederates had raised *Merrimack* and floated her into the dry dock at Gosport, that Welles decided to step up the ironclad campaign. He needed Congress to pass a bill authorizing the creation of an ironclad board that could advertise for, and examine, proposals for ironclad vessels, and then authorize their construction. To make certain the bill would say what he wanted it to say, Welles wrote it himself. But he needed someone else to lobby for it.

Welles found just the man.

A Railroad Man from Connecticut

As it happened, the man Welles approached to bring his ironclad bill to Congress would ultimately become the prime mover in the creation of the *Monitor*, an advocate without whom it is likely that that groundbreaking ship would ever have been built. That man was Cornelius Scranton Bushnell, a well-connected entrepreneur, speculator, and president of a small Connecticut railroad.

Cornelius Bushnell was only thirty-one years old at the commencement of the Civil War. His family had been in New Haven, Connecticut, since it was New Haven Colony, the first Bushnell settling there in 1638. Cornelius himself was from Madison, about twenty miles east.

Despite the family's pedigree, Cornelius Bushnell was born into poverty. At fifteen, the ambitious Cornelius went to sea as an apprentice aboard a coasting vessel. A year later he was master of a sixty-ton schooner. Even as a teenager, Bushnell was displaying the kind of drive and ambition that would ultimately make him a wealthy and influential man.

In the spring of 1858, Bushnell and his brother Nathan were building their fortunes as grocery merchants. On hearing that the New Haven & New London Railroad was on the verge of liquidation, Bushnell purchased a single share of stock, giving him access to the share-holders' meeting. There he convinced his fellow shareholders that rather than abandon the railroad, they should form a committee to se-cure continuing financing and even extend the line from New London to Stonington.

When the expanding railroad once again ran out of capital and the president resigned in despair, Bushnell convinced the board of direc-tors to appoint him to the post to, as he put it, "see what a boy could do." That particular boy was able to turn the railroad around, making it a profitable enterprise. In a little more than a year, Bushnell had gone from owning a single share of stock to being elected president of the New Haven & New London Railroad.

By then Bushnell was already well known in Connecticut's political circles. His work on the railroad, including securing government con-tracts to carry the mail, now brought him to Washington and intro-duced him to a number of important figures on the national stage. It was those connections that would make Bushnell the pivotal player in the birth of the *Monitor*.

Through 1860 and early 1861, Bushnell spent considerable time in the nation's capital and witnessed a great deal of the political history leading up to the Civil War. He heard Jefferson Davis deliver his farewell address to the Senate, and most of the other United States senators from seceding states as well, leaving to take their places in the Confederacy.

On Monday, March 4, 1861, Bushnell once again traveled to Wash-ington, D.C. on railroad business. He took a room at the famous Willard's Hotel, where anyone of importance stayed while in Washing-ton, and where a good deal of the nation's business was transacted. Also at Willard's that day was President-elect Abraham Lincoln, meeting with William Seward. Lincoln had come to convince Seward, his rival for the presidency, to remain as part of the cabinet. Soldiers and artillery were posted around the Capitol. It was the day of Lincoln's inauguration.

Nationwide, tensions were quickly approaching a breaking point, with seven states having seceded and more sure to follow. The possibil-ity of civil war, long considered inconceivable, was becoming very con-ceivable indeed. Along with his mail contracts, Bushnell was looking for a way to participate in the coming conflict. He would find it sooner than he expected.

Bushnell was still at Willard's when the war's first shots were fired. On April 12, 1861, General Pierre Gustave Toutant Beauregard ordered the batteries scattered around Charleston Harbor to open fire on the small Union garrison in Fort Sumter. The Civil War had begun and, as Bushnell said after, "the people of the North awoke as from a dream."

Dream became nightmare in Washington, D.C., where it was supposed by the Unionists that more than half of the population were Southern sympathizers, and that the city was threatened from within. Bushnell and the other Northern-leaning men at Willard's surreptitiously gathered in the hotel's parlors, where General James Nye rallied them to the fight.

"Tonight we are threatened," Nye told them, "with an invading force from over the Potomac, to unite with the sympathizers in this city, determined on taking possession of this government. All who are willing, if need be, to become the first victims of this causeless and wicked rebellion, will repair at once to the church in the rear of the hotel and enroll in the Cassius M. Clay battalion for the defense of the City of Washington."

Cassius M. Clay was an abolitionist and a two-fisted politician. In the days before attack ads, Clay was known for simply attacking. Once, while campaigning, he killed a political opponent with a Bowie knife. Lincoln had appointed him minister to Russia, but Clay delayed his departure to organize his Ninety Day Guard for the defense of the capital. Thus, on the first night of the Civil War, young Cornelius Bushnell found himself prowling the streets of Washington, D.C. with a newly issued Springfield rifle in his hands, looking for invading Confederates and fifth columnists.

Despite the general fear and Nye's high-blown rhetoric, they didn't find any. Soon after Clay's troops were mobilized, the 6th Massachusetts arrived to garrison the capital. Though Nye had called for men willing to be the "first victims" of the conflict, it was actually men of the 6th Massachusetts who earned that distinction. While marching through Baltimore, the 6th had had a run-in with a mob that led to gunfire and left four soldiers and twelve Baltimore civilians dead.

Bushnell and his fellow irregulars were greatly relieved to see the 6th, as was all of Unionist Washington.

Bushnell's Civil War began with a moment of high drama, but the young capitalist's true talents were not in shouldering a weapon but in lobbying and dealmaking. Among his numerous contacts in the capital was Commodore Hiram Paulding. Upon returning from the disaster at Gosport, Paulding met Bushnell and suggested a new line for the railroad man—ironclads. Though there was of yet no government call for such vessels, Paulding suggested Bushnell "devote [his] efforts to the construction of iron batteries."*

Cornelius Bushnell was clearly a man who could spot opportunity, and he saw it in Paulding's suggestion. In Bushnell there was the confluence of patriotism and business, and his efforts were directed at projects that would be of benefit to both his country and his and his associates' bottom line. With those twin motivators, he entered the world of ironclads.

Bushnell had been a shipmaster on a small scale, but he was not a ship designer. Soon after speaking with Paulding, he hired "one of the ablest young constructors from Boston," Samuel H. Pook.

The Civil War did not start well for Pook. As the senior naval constructor in New York, he was one of the men charged by Welles to purchase merchant ships for the blockading fleet. Pook was taken to the cleaners by New York shipowners. It was only after Welles received reports of similar transactions from Boston that the navy secretary realized the problem went beyond one man.

Pook was already known as a brilliant naval architect, which was fortunate, because he would not have gained that reputation based on the work he did for Bushnell.

Pook and Bushnell went to work on creating their most innovative ironclad design. In the end, it was not terribly innovative at all. *Galena*, as she would be called, was a schooner-rigged, propeller-driven steamer, 210 feet long, built of wood. She was a very conventional ship with iron bars attached lengthwise to her bulging topsides.

The iron would prove to be as much a danger as a help. After an

---

*The use of the term "iron batteries" as opposed to "ironclad ships" or "ironclad menof-war," and the constant interchanging of terminology, indicates just how new and untried this revolutionary technology was. It was still not certain what these things were—ships, floating batteries, self-propelled floating batteries? It was an ambiguity that would last for almost a year, until the *Monitor* and *Virginia* settled it for good.

encounter with a Confederate shore battery, Commander John Rogers wrote of *Galena*, "She is not shot-proof; balls came through, and many men were killed with fragments of her own iron. One fairly penetrated just above the water line and exploded in the steerage." After three months of design work, she was, for Bushnell, "the best result that I had been able to attain."

For all her shortcomings, *Galena* did good service in the navy, which included taking part in Farragut's attack on Mobile Bay. Still, Cornelius Bushnell was about to learn what real design innovation was all about.

In early July, Bushnell was once again in Washington, once again staying at Willard's Hotel. With the design phase of his ironclad nearly complete, it was not surprising that he should pay a visit to his old associate and fellow Connecticut native Gideon Welles, the man who could most help him further his project, this despite the fact that the U.S. navy had not yet taken any formal interest in ironclad ships.

The two men sat down to a discussion of ironclads and the ongoing conversion of *Merrimack*. Welles complained to Bushnell that he had "called the attention of Congress" to the need for ironclads, but "without effect."

Perhaps it was at that moment that Welles realized he was talking to just the man he needed to help get a bill through Congress. Bushnell was a seasoned lobbyist, and he had a vested interest in seeing this legislation pass, as it would appropriate funds for building ships just like the one he was developing. The congressman from Bushnell's district, James E. English, with whom Bushnell had a working relationship, sat on the House Naval Committee.

Welles asked Bushnell if he would take on the job of getting through Congress a bill that he, Welles, would write. Bushnell agreed. That same night Welles met Bushnell at Willard's and handed him a copy of the bill. Bushnell went to work.

His first call was to James E. English. The congressman was enthusiastic about the idea and brought it to the attention of Charles Sedgwick of New York, chairman of the House Naval Committee, who was also amenable. The three of them worked on the other members, as well as those who sat on the Senate Committee for Naval Affairs. Their lobbying effort took about two weeks, fast by congressional standards.

On August 3, Congress approved "An Act to provide for the construction of one or more Floating Batteries." The act read:

> . . . the Secretary of the Navy be, and he is hereby, authorized and directed to appoint a board of three skilful naval officers to investigate the plans and specifications that may be submitted for the construction of iron or steel-clad steam ships or steam-batteries, and on their report, should it be favorable, the Secretary of the Navy will cause one or more armored or iron or steel-clad steamships or floating steam batteries to be built; and there is hereby appropriated . . . the sum of one million five hundred thousand dollars.

The next act Congress voted on provided for a twenty-dollar fine for anyone selling liquor in the District of Columbia.

That same night, Lincoln signed the bill into law. The Union was officially in the business of ironclads.

Chapter 10

An Ironclad for the South

Before construction of a Confederate ironclad could begin, Brooke's rough drawings had to be turned into plans from which a ship could be built.

Mallory sent to the Gosport Naval Shipyard for someone who could do that work, and a master ship's carpenter was sent to Richmond. But the carpenter was not a draftsman either, and was not very helpful, "lacking in confidence and energy, and was adverse to performing unusual duty." He helped Brooke with the specifications for the wooden part of the vessel, but could do little else. Soon he was complaining that the water in Richmond made him sick. He was permitted to leave.

In his stead, Mallory ordered John Luke Porter to come to Richmond. Porter was a constructor—the contemporary term for a naval architect—and had been working with iron vessels as far back as 1846. In 1857, Porter became a constructor for the U.S. navy, having failed an examination for the post ten years earlier. Porter was a native of Portsmouth, Virginia, and was stationed at the shipyard when the Union abandoned the facility. He resigned from the U.S. Navy and immediately secured a position with the Confederacy as a constructor, the only U.S. Navy constructor to do so. As one of his first duties, he helped draw up the estimate of the value of what was left behind when Norfolk was abandoned.

Like many of those involved in man-of-war construction, Porter had long been interested in ironclads. While supervising the construction of an iron vessel in Pittsburgh in 1846 he had come up with a design for an ironclad and built a model of it. He presented the model to the Navy Department, but the navy was not interested. They were already a few years into the experimental Stevens Iron Battery, which was expense and headache enough. Porter put his ironclad ideas aside.

With the outbreak of the Civil War, Porter, like Mallory and Brooke, understood that the Confederacy would need ironclads to match the Union's naval superiority. When Mallory summoned him to Richmond, Porter built a new model, just like his original, and took it with him.

Also called to Richmond to discuss Brooke's plans was Chief Engineer William P. Williamson. Williamson had entered the engineering branch of the navy shortly after its creation in 1842, one of the first officers in that department. His first assignment was to the ship *Union*, a bizarre experimental steamer that mounted two paddle wheels horizontally below the waterline. Williamson had assisted the inventor, Lt. W.W. Hunter, on the prototype, *Germ*. Now, with Hunter in command, *Union* sailed on a test voyage. It was not a success. Six years later, *Union* was made a receiving ship in Philadelphia.

Williamson, too, had an interest in ironclads. Like Porter and Brooke, he had made drawings of his own ironclad concept three years before.

By the outbreak of the war, Williamson had been stationed at the Gosport Naval Shipyard for five years. A native of North Carolina, he left the U.S. Navy and continued on in his former capacity for the Confederate Navy. In June 1861 he traveled with Porter to Richmond.

The four men, Porter, Williamson, Brooke, and Mallory, met in Mallory's office. There they examined the model and the drawings Porter had brought with him. Not surprisingly, Brooke fully approved of Porter's casemate design, since Porter's and Brooke's "were almost identical." This was not as much of a coincidence as it might seem, since the concept of a floating ironclad battery with inclined sides was well established. Like Brooke's casemate, Porter's iron plating extended to 2 feet below the waterline.

Mallory then asked Porter and Williamson to examine the drawings that Brooke had made. They showed a vessel much like Porter's model, of course, except that the hull extended far out in front and in back of the casemate.

This in fact was the major difference between the two designs. Porter's "consisted of a shield and hull," as Brooke later described it, "the extremities of the hull terminating with a shield, forming a sort of box or scow, upon which the shield was supported." Porter's ironclad was "not calculated to have much speed, but . . . intended for harbor defense only," Porter admitted.

Brooke explained his design to Porter and Williamson, showing how the fine lines and extended hull would give his vessel more buoyancy and speed. The two men agreed that Brooke's plan was the better of the two. Or so Brooke claimed.

At the time, Porter apparently expressed no objections to Brooke's idea. That night Brooke wrote in his journal, "I was afraid Mr. P_____ would, having an idea of his own, make objection to my plan but he did not regarding it as an improvement."

It was some time later, after *Virginia* proved a success and Brooke began receiving public credit for the design, that Porter began to object. He made the claim that he was the one who designed *Virginia*, that Brooke had not come up with any design or shown any drawings at the meeting.

The first public acknowledgment of Brooke's having designed the ship appeared in the *Richmond Enquirer and Whig* in an article written, according to Porter, by an officer in Brooke's office and at Brooke's direction under the *nom de plume* "Justice." Porter claimed that Williamson had warned him that Brooke had a propensity to "appropriate other people's plans for his own."

Brooke in turn would insinuate that the first ship's carpenter whom Mallory had summoned to Richmond had, upon returning to Portsmouth, told Porter of the design Brooke had worked out. He claimed Porter built his model based on that intelligence before meeting with Mallory. The fight, played out in various publications, was not pretty.

Though Williamson did not jump into the fray, his son, who later became a chief engineer in the U.S. navy, claimed that the drawings his father had made of an ironclad were exactly what the reconstructed

Merrimack became. It was his opinion that his father had submitted the plans to Porter and Porter had incorporated the engineer's ideas with his own. Including at least two others on record as having designed the *Virginia*, the number of claimants for that honor stands at five.

Porter's claims were not without merit, but the only disinterested party involved, Steven Mallory, always maintained that Brooke was the one who had originated the design, and that would appear to be the most likely case. Still, the argument between Porter and Brooke over who designed the famous ironclad would last for years, would last, in fact, until they were both dead.

The Only Option

Regardless of whose idea it was, Mallory now had in front of him a workable concept for an ironclad, and he did not want to waste a moment. He instructed Porter and Williamson to find out if a suitable engine and boilers could be had. He told Brooke, "Make a clean drawing in ink of your plan, to be filed with the department."

Brooke was spreading the paper out on the table when Porter interrupted him. "You had better let me do that," Porter said. "I am more familiar than you are with that sort of work."

Porter began to draw the proposed ship—his drawings were for reference only, not actual working plans—while Brooke and Williamson looked into the question of engines. They went first to the largest iron foundry in the Confederacy, Anderson and Company, better known by its old name, the Tredegar Iron Works. Tredegar was about a mile away from the Navy Department office on the banks of the James River. If suitable engines were to be found anywhere in the South, it was there. But Tredegar had no engines.

Brooke and Williamson despaired of finding any anywhere else in the Confederacy. Then Williamson had another idea.

On May 20, the commandant of the Gosport Naval Shipyard, French Forrest, had written to the owners of a local salvage company, B & J Baker and Company, "You can begin under your contract to raise the 'Merrimack.'"

All of the resources of the Gosport Naval Shipyard—and despite the destruction carried out by the retreating Yankees, they were considerable—were used to assist in raising the ship and towing her into dry dock. The salvage work cost six thousand dollars.

A few days before Mallory arrived in Richmond, Forrest sent a short note to Robert E. Lee: "We have the 'Merrimack' up and just pulling her in the dry dock."

If the federals had simply burned *Merrimack* as she floated on her waterline, and not scuttled her first, there would have been nothing for the Confederates to salvage. But as it was, the water flooding the hull protected the lower part of the vessel from the flames, and left it virtually intact. Williamson, as chief engineer at the yard, had examined the engines and boilers and was well acquainted with their condition. He suggested that the engines could be put into working order, and what was left of the steam frigate be turned into an ironclad.

This suggestion was not ideal. The plans that Porter was drawing up, based on the men's shared vision, called for a hull of about 8 feet draft, perfect for the shallow water in Southern harbors. The frigate *Merrimack* had drawn 24 feet aft, and an ironclad built on her hull would not draw much less. Any hull in which the *Merrimack*'s engines were installed would have to be at least as deep as the frigate's.

On the other hand, this plan did offer a big savings in time and money. Since the old hull was intact, it was pointless to build another. The engines and boilers were already in place. And just as important, the propeller shaft, which would prove a big obstacle in the construction of other ironclads, was there as well, as was the propeller. Captain Samuel Barron had estimated that the partially burned *Merrimack* was worth two hundred and fifty thousand dollars. Using *Merrimack*'s hull, half the ship would be built and paid for before they even started construction.

Porter, who was familiar with the condition of *Monitor*'s hull, agreed. The ironclad casemate could be built on top of the existing hull. Though not delighted by the idea, the three men recognized that they had no other choice. They reported their plan to Mallory.

After some discussion, Mallory agreed. For the record, he asked that the three men write up a report based on their discussion. The report, submitted on June 25, began:

*In obedience to your order we have carefully examined and consid-
ered the various plans and propositions for constructing a shot-proof
steam battery and respectfully report that in our opinion the steam
frigate* Merrimack, *which is in such condition from the effects of fire
as to be useless for any other purpose without incurring a very heavy
expense in her rebuilding, etc., can be made an efficient vessel of
that character . . .*

They went on to point out that the "bottom of the hull, boilers and
costly parts of the engine" had suffered little damage. The casemate built
on top of the hull would be the same as the one on which they had already agreed. They estimated the work would cost one hundred and ten
thousand dollars, with most of that money going toward labor. The materials they needed, save for iron plate, had been left behind by the federals.

The three men divided the labor among themselves. Porter was to
draw up the working plans of the ironclad. Williamson would be responsible for rehabilitating the engines. Brooke would oversee the manufacturing of the iron plate and determine the best configuration of
wood and iron for the shield.

Porter returned immediately to Norfolk and began to draw the plans,
modifying the original vision to accommodate the reality of the *Merrimack*'s hull. Since Porter was doing the drawing, he drew it the way he
wanted. "I . . . placed the very same shield on her," Porter wrote, "which
was on the model I carried up with me . . ." With the shield going on the
old frigate's hull, the ironclad would have a streamlined shape and an
extended bow and stern whether Brooke had designed it that way or not.

Since Porter's shield looked much like Brooke's, Brooke was not
even aware that Porter had put his own design on the plans. Brooke later
testified before Congress that Porter "sent up drawings which were of
the same general description as those he made before in accordance
with my suggestion." Both men thought they had designed the ironclad.

While Porter was making the plans, Mallory was asking for the
money. In his July report to President Davis, Mallory estimated the
cost of returning the *Merrimack* to her former condition, as a sail and
steam frigate, would be around four hundred and fifty thousand dollars. "The vessel would then be in the river," he pointed out, "and by the

blockade of the enemy's fleets and batteries rendered comparatively useless."

Fortunately, there was a better way. For only $172,523 (a new esti- mate worked out by Porter and Williamson), it was possible to "shield her completely with 3-inch iron, placed at such angles as to render her shot proof . . ." The work would be completed as quickly as possible, the ship armed with the heaviest ordnance and sent against the enemy. As time was of the essence, Mallory had not hesitated in ordering the work commenced and asking Congress for the money. A month later Congress approved the request.

By July 11, Porter had completed the plans for converting the *Mer- rimack* into an ironclad. He carried them to Richmond and delivered them to Mallory. Mallory then wrote an order to French Forrest and handed it to Porter for the constructor to personally deliver:

> Sir:
> *You will proceed with all practicable dispatch to make the changes in the* Merrimack *and to build, equip and fit her in all respects ac- cording to the designs and plans of the constructor, and engineer, Messrs. Porter and Williamson.* As time is of the first importance in the matter, you will see that work progresses without delay to com- pletion.*

There would be delays, of course, frustrating and heartbreaking de- lays, due mainly to the Confederacy's want of industrial capacity and inadequate railroads. But as of July 11, the Confederate navy began construction of an ironclad, a secret weapon that they hoped would sweep the seas of the Yankee invader.

---

*Porter would later use this wording as further evidence that he was indeed the *Vir- ginia*'s designer. In truth, since Brooke would have nothing to do with the actual recon- struction, and would not even be at Norfolk, there was no reason for Mallory to include his name in the orders to Forrest.

Chapter 11

"Three Skilful Naval Officers"

On August 7, 1861, the United States government issued contracts for the construction of seven ironclad vessels. Ironically, they had nothing to do with Gideon Welles or the navy department.

The contracts were issued by the War Department—the army—to Mr. James B. Eads of St. Louis. The ironclads were the seven City class gunboats designed by Samuel M. Pook (not to be confused with Samuel H. Pook, who had recently completed drawing *Galena* for Cornelius Bushnell). The City class boats, intended for fighting on Western rivers, would have navy men for officers. But for the first year of their lives, they were decidedly an army affair.

To further the irony, August 7 was also the day that the Navy Department, lagging behind their shore-based brethren in arms, issued advertisements calling for plans and cost estimates for ironclad vessels, built either from iron or wood and iron combined.

The advertisements were fairly specific as to what the navy was looking for: ships that could carry armament weighing from eighty to 120 tons, that could hold provisions for sixty days and coal for eight days' steaming. Recognizing that such ships would be most useful inshore, the specifications called for a draft of not less than 10 feet but not more than 16, with the shallower draft preferred. Still hanging on to what they knew best, the Navy Department also called for each ship to have at least two masts with wire standing rigging. Plans were to be submitted within twenty-five days of the date of the advertisement.

The Ironclad Board

The following day, Gideon Welles appointed the "three skilful naval officers" who would make up the ironclad board. The senior officer and chairman of the board was Commodore Joseph Smith, a fifty-two-year veteran of the U.S. Navy. Smith was born just eight years after the end of the American Revolution. He entered the navy in 1809, was wounded at the battle of Plattsburgh during the War of 1812, and served aboard USS *Constitution* during the war with Algiers.

In 1846 he became chief of the Bureau of Yards and Docks, the same year that Welles was named chief of the Bureau of Provision and Clothing. The two men developed a friendship and mutual trust that lasted through the years, and continued on after Welles returned to Washington fifteen years later as Smith's civilian boss.

Smith was still chief of the Bureau of Yards and Docks when Welles tapped him for the ironclad board. Smith would have the responsibility of overseeing the construction of any vessels the board decided to build. Despite his long tenure in the sailing navy, Smith was perhaps the most open-minded of the three board members, and the most amenable to the idea of ironclad ships.

Joining Smith was Commodore Hiram Paulding, fresh from his adventures at the Gosport Naval Shipyard. The third appointee was Commander John Dahlgren, the ordnance expert, but at his own request he was relieved of that duty. Commander Charles Davis took his place. The three men, skillful though they may have been, had no illusions as to their expertise in the matter of ironclads. As a committee they would write, "Distrustful of our ability to discharge this duty, which the law requires should be performed by three skillful naval officers, we approach the subject with diffidence, having no experience and but scanty knowledge in this branch of naval architecture." They did, at least, know how to spell "skillful."

The board applied to the Navy Department to have a naval constructor assigned to them to offer an expert opinion, but with all of the construction going on there were none to be had. The three officers were on their own, and they did the best they could.

Tradition-bound and ignorant of ironclad technology though they may have been, the board was not necessarily adverse to the concept. "For rivers and harbor service we consider iron-clad vessels . . . as very important," they reported. They felt ironclads could be of use in running past fortifications or in reducing temporary batteries, but they did not believe that any ironclad was a match for a well-built, land-based fortification. They adhered to the age-old belief, which was not entirely wrong, that gunners aboard a floating, moving battery could not lay their guns as well as gunners on the stable platform of a fortress. And an ironclad, they felt, simply could not carry enough armor to match the walls of a fort and still float.

Further, the board did not believe ironclads would be useful as anything beyond harbor defense or river work. "As cruising vessels . . . we are skeptical as to their advantage and ultimate adoption." The board felt that the weight of the iron would necessitate so wide a beam as to make the ship slow, unwieldy, and a great consumer of coal. While it is tempting to see the three officers as shortsighted, in light of the technology of the time, their assessment was actually quite well reasoned and correct. Their concerns about seaworthiness, dismissed by ironclad proponents as the irrational fear of the ignorant, would prove prophetic.

The board correctly saw ironclad design falling into two broad categories: fairly traditional sail-and-steam ships with some form of iron shell over a wooden hull—the *La Gloire* model—and the "floating battery"—low, boxy, raftlike vessels for inshore work. *Galena* represented the former type. The vessel that the Confederates were building out of the old *Merrimack* was of the latter.

By the end of the one-month period allotted to submit plans, the board had received proposals for sixteen ironclad designs. Actually, there were fifteen ironclads and one rubberclad, submitted by William Kingsley of Washington, D.C. The board did not recommend its construction.

Perhaps because a month was not a very long time to design an ironclad from the ground up, a number of the proposals lacked information. Such was the case with Henry R. Dunham of New York, who submitted "no drawings or specifications," save for the general dimensions of the ship, as well as the cost: one million two hundred thousand dollars. The board passed.

Almost as expensive as Dunham's ship was one proposed by Donald McKay, the famed clipper ship designer from Boston. A decade before, McKay had designed and built *Flying Cloud*, the ship that broke the New York to San Francisco record during her maiden voyage. The board liked what they saw of McKay's design, but the one-million-dollar sticker price was too much for them, and they passed on that one as well. (McKay did not take it well. He was vocal in his criticism of the Navy Department and ridiculed the *Monitor* in the press shortly before her appearance at Hampton Roads.)

One of the board's chief concerns, a concern shared by many sailors in regard to ironclads, was the issue of stability. With all the weight of the armor plating above the waterline, it was feared that ironclads would either be too unstable to be safe or would require so broad a beam that they would be unmanageable. One design was rejected with the words, "plan good enough but breadth not enough to bear the armor." Another was dismissed with "Dimensions of vessel, we think, will not bear the weight and possess stability."

The Almost Final Decision

Near the end of the review process, the ironclad board was ready to recommend two vessels. One was a 220-foot-long, 60-foot-wide wood-and-iron ship proposed by Merrick and Sons of Philadelphia. When first built she was rigged as a bark, but the sailing gear was eventually removed, leaving her a pure steam vessel. She was launched a little less than a year after approval, and carried a powerful battery of two 150-pound Parrott rifles, two 50-pound Dahlgren rifles, fourteen 11-inch Dahlgren smoothbores, one 12-pound rifle, and one heavy 12-pound smoothbore. She was christened *New Ironsides* and would serve her country well during the war.

The other vessel approved was the *Galena*. The forward-thinking

Cornelius Bushnell had been working for months on his plan by the time he showed it to the board, so there was no issue of an incomplete proposal. Bushnell was so confident that his ship would be approved that he arranged the subcontracting of the iron plate with two ironworks in Troy, New York, before he even presented his plans to the board.

Despite Bushnell's confidence, the ironclad board gave only qualified approval of *Galena*. They reported that their "objection to this vessel is the fear that she will not float her armor and load sufficiently high, and have stability enough for a sea vessel. With a guarantee that she shall do these, we recommend on that basis a contract." It was the old bugaboo of stability with iron carried so high above the waterline.

The ironclad board would not take the word of Bushnell or Samuel Pook that *Galena* had sufficient stability. They wanted mathematical proof that she was safe, but that Bushnell did not know how to provide. He left the meeting with one more hoop to jump through.

Back at Willard's Hotel, that epicenter of business and politics, Bushnell ran into Cornelius Delamater, co-owner of Delamater & Company, a major ironworks in New York City. Delameter was in town to see Gideon Welles and make his ironworks available for any government contracts that might need filling.

The meeting of those two men, Bushnell and Delamater, was one of those unique instances, a moment of no obvious significance to the participants that would, in the end, change the course of history. Bushnell confided his problem to Delameter. Delameter knew just the man to prove *Galena*'s stability, a man "whose opinion would settle the matter definitely and with accuracy": John Ericsson of New York.

The next day Bushnell boarded a train heading north. He arrived in the late afternoon at Ericsson's residence, 95 Franklin Street in New York City, a home the expatriate Swede had occupied since 1843, three years after arriving in America by way of England.

Ericsson's longtime devoted house girl, Ann Cassidy, ushered Bushnell into the house and led him to the room used by Ericsson as his office. The man who greeted Bushnell was fifty-eight years old, five foot seven and a half inches tall, with a military bearing. He had a large, firm-set jaw flanked by bushy sideburns that gave him a no-nonsense look, which was quite appropriate, because John Ericsson was a no-

nonsense man. He was neat and careful about his appearance and dressed well, though in no way lavishly.

Ericsson welcomed the travel-weary Bushnell into his office. Bushnell explained to Ericsson why he had come. He showed Ericsson *Galena's* plans and the data that the engineer would need to make the stability calculations. Ericsson looked them over. He told Bushnell to come back the next day, and he would have an answer. Bushnell thanked him and left.

That first meeting, as it turned out, was the beginning of a lifelong relationship between the two men. Along with their partnership in one of the most important moments in naval history, as well as other ventures, the two men would form a deep friendship. Ericsson would name Cornelius Bushnell executor of his estate. Bushnell would name his son Erricsson F. Bushnell. (Ericsson, with his usual propensity for exactness, would one day write to Bushnell's son, "Allow me to call attention to the fact that your name should be spelled with a single r.")

The next day, Bushnell returned to the house on Franklin Street. Ericsson showed him an elaborate series of calculations that must have been as clear as Chinese to the railroad man. But Ericsson gave him the good news in plain English. "She will easily carry the load you propose," he said, "and stand a six-inch shot at a respectable distance."

Though enduring "a six-inch shot at a respectable distance" is hardly unqualified praise—*Virginia*, the ship *Galena* was ostensibly built to defend against, would have ripped her apart with her 9-inch Dahlgrens at point-blank range—still, it was enough for Bushnell. The ironclad board, after all, was worried about stability, not resistance to shot.

As their meeting drew to a close, Bushnell made to leave, but Ericsson stopped him. He asked Bushnell if he would like to see his, Ericsson's, own plan for a floating battery, "absolutely impervious to the heaviest shot or shell."

Ericsson was generally hesitant about showing his plans and models. A former friend and employee who had worked for years as Ericsson's assistant wrote, "He was very reserved about his models and inventions and seemed to have a mortal dread of their being discovered." But Ericsson must have seen something in the ambitious young Bushnell that he liked. The inventor pulled out a dusty box and opened it up. Inside was a pasteboard model of a vessel the likes of which Bushnell had never seen.

The vessel had a deck that was roughly the shape of a football. In cross-section it was vaguely triangular in shape, with the apex of the triangle forming the bottom of the ship. It was a model of what would become the USS *Monitor*. Almost.

The model, and the plans that went with it, were developed by Ericsson in 1854. At the time, France and England were allied against Russia in the Crimean War. Ericsson, a native of Sweden and as such a born enemy of Russia, was eager to help Russia's enemies. Ericsson sent the plans and the model to Emperor Napoleon III of France. Napoleon was apparently impressed. He sent Ericsson a letter of thanks and even a medal for his efforts, which Ericsson proudly displayed for Bushnell. But the Swede's "new system of naval attack" came too late in the war and was deemed too expensive to be worth pursuing. The French did not act on it.

If the model Ericsson showed Bushnell was in fact a model based on the plans he developed for Napoleon, then there were several differences between it and the ship that became the *Monitor*. Rather than the essentially flat deck of the *Monitor*, the deck of the original design was rounded, and the point where the deck met the hull was underwater. Floating on her waterline, the vessel would have looked very much like a modern submarine on the surface.

The original version of the ship did not have the pilothouse of the later vessel. Instead, those who needed to see out could look through "small holes at appropriate places." She was also intended to mount what were essentially periscopes.

Perhaps the greatest difference was the turret. Rather than the famous "cheese box" shape of the *Monitor*'s turret, the original turret was a dome, like a cupola. In his letter that accompanied the model, Ericsson made much of the dome's properties, claiming, "Shot (of cast iron) striking the globular turret will crumble to pieces or are deflected."

"Alas! for the wooden walls that formerly ruled the waves!" Ericsson wrote to Napoleon. But as it happened, the wooden walls would get an eight-year reprieve before iron would usurp their crown.

It is not clear why the final version of Ericsson's ironclad differed from his original vision. Perhaps the added difficulty of manufacturing the complex curves needed for the sloping deck and globular turret made it impractical when cost and speed of manufacture were vital considerations.

Whatever the reason, Bushnell was immediately enamored of the model and the plans. After Ericsson had lectured for ten minutes concerning the vessel's advantages, Bushnell "awoke to the fact that salvation was in store for our Government and country." Though that might have been a bit hyperbolic—hyperbole attended both *Monitor* and *Virginia* for most of their careers—Bushnell was correct in recognizing the importance of this ship. After all, he and Pook had been working for months and could come up with nothing better than putting iron plate on a conventional sail-and-steam vessel. Here was something revolutionary.

Ericsson felt that his ship was vital to the security of the United States, and he had already made an effort to bring it to the government's attention. Eschewing the navy, he went right to the top, writing to President Lincoln on August 29, perhaps a day or so before Bushnell's visit. Ericsson intended for the battery to steam to Norfolk and destroy the ships left behind by the Union and rout the Confederates. Ericsson laid out his qualifications, mentioning, with his usual humility, that he possessed "practical and constructive skill shared by no engineer now living."

He further stated his reasons for offering the ship to the United States, which were not pecuniary but rather a strong love of the Union. The letter was never acted on. But after Bushnell became involved, events began to move so swiftly that efforts to build the ship were under way possibly before the letter even made it to Lincoln's desk.

Bushnell asked Ericsson if Ericsson would lend him the model and the plans, so that he might present them to the ironclad board. Though Bushnell might have seen an opportunity for another piece of the ironclad pie—which *Monitor* would end up being for him—still, there was a strong patriotic motive in place. After all, he and Ericsson would be competing for the same government funding. There was no set number of ships the ironclad board needed to approve, and approval for *Galena* was only provisional. It was possible that the board might forgo building *Galena* in favor of Ericsson's battery, and Bushnell would have talked himself out of a contract. But Bushnell, like Ericsson, genuinely saw the benefit that this floating battery could provide to the nation they both loved.

Getting others to see the benefit would be the problem. Ericsson was already a somewhat famous man, and Bushnell probably knew quite a bit about his history with the navy. No doubt Bushnell knew that Ericsson loathed the U.S. Navy, and that the navy felt likewise, and that the navy still owed John Ericsson more than $15,000, money they refused to pay and indeed never did.

Bushnell certainly knew that the navy still considered Ericsson partially responsible for the deaths of six men, including President Tyler's secretaries of state and the navy.

Chapter 12

Merrimack Redux

In July 1861, John Luke Porter returned to the Gosport Naval Shipyard bearing Mallory's order to Forrest to commence work on the conversion of the *Merrimack*. The ship was "burned to her copper line, and down through to her berth deck, which, with her spar and gun decks, were also burned." She had been sitting in the dry dock for two weeks waiting for a decision as to her fate.

Porter got right to work, "unassisted by mortal man so far as the plans and responsibilities of the hull and its workings were concerned." And it is certainly true that Porter bore nearly all of the responsibility for turning the burned-out wreck into a fighting machine. Williamson had only to refurbish the engines already in place. Brooke was coordinating with Tredegar for iron plate and experimenting with ordnance. He only rarely came to Gosport (Porter said only once, but it was certainly more than that). The real bulk of the work—design and implementation—fell to Porter. It is little wonder that he was resentful of Brooke getting the credit.

Porter wanted to cut the existing hull down as far as he could, to give the ship the absolute minimum draft. He calculated all of the weight that would go into her, the weight of the iron-and-wood shield, the engines and boilers, guns, shot, stores, and coal. He found he could cut the hull down to 19 feet from keel to gunwale and that would still give him an extra 50 tons of displacement.

Unfortunately, when he struck the line on the side of the ship, he found that 19 feet would leave the top of the propeller a foot out of the water. He could cut the propeller down, but that was time-consuming and would have reduced her speed, and Porter might have guessed that she would not have speed to spare. He raised the line to 20 feet aft.

Porter appointed ship's carpenter James Meads, formerly of the U.S. navy, as the yard's master ship carpenter. Meads had assisted Porter in assessing the damage done to the Union ships at the navy yard. Now he served as master carpenter on the *Merrimack* project. Meads would supervise the thousand to fifteen hundred men who labored on *Merrimack* from July until her launch as CSS *Virginia* seven months later.

As soon as the *Merrimack*'s hull was cut down to the line Porter had determined and all the charred wood cleared away, work began on the shield, or casemate. Oak knees cut at a 35-degree angle were fitted between the original frames of the ship and bolted in place. Yellow pine beams, 16 inches thick, which Porter referred to as "rafters," formed the sides of the casemate. These were bolted to the knees and ran vertically up from the edge of the existing hull of the ship, forming casemate sides that sloped up at a 35-degree angle from the horizontal.

At the front and back end of the casemate, the rafters were arranged to form rounded ends with the same 35-degree angle. Like the framing of a wooden man-of-war, virtually no space was left between the rafters, making them "solid fore and aft the whole length of the shield."

The angle formed by the juncture of shield and hull, where the pine rafters met the original frames of *Merrimack*, Porter called the "knuckle." The ship was intended to float with the knuckle 2 feet below the water, thus protecting the vulnerable shield-to-hull joint.

In keeping with traditional wooden shipbuilding techniques, Porter installed inner planking, or "ceiling planks," on the inside of the rafters. These ran horizontally and were also made of yellow pine, 4 inches thick.

At this point, Brooke apparently made one of what would be a number of suggested changes that would drive Porter to distraction. Brooke suggested removing the ceiling planking and placing it on the outside of the shield as backing for the iron plating, which ultimately was done.

On top of the pine planks, which ran lengthwise, oak planks 4 inches thick were attached vertically. In all, the iron plates would be backed up by 24 inches of wood.

Guns and Iron

On July 23, the capital of Richmond was still wild with jubilation. Two days before, the Confederate Army had routed the Union forces at Manassas and proved what most "Southrons" already knew—that Yankee shopkeepers were no match for Southern boys on the battlefield.

That same day (and a month before the Confederate congress had authorized the money for it), Mallory entered into a contract with the Tredegar Iron Works for the manufacture of the *Virginia*'s iron plate, as well as her other iron needs. The price negotiated was six and a half cents per pound. It was estimated that the ship would need 1,000 tons.

The first job was finding the iron. Fortunately, quite a bit had been liberated from the North by General Thomas J. Jackson, who had just earned his nickname of Stonewall. Jackson's troops had and continued to systematically rip up the tracks of the Baltimore and Ohio Railroad in the Shenandoah Valley to disrupt Union communications. Now the War Department ordered that all that rail be turned over to Tredegar.

More iron came from other Southern railroads, abandoned as being too exposed to enemy attack, such as the rail from Winchester to Harper's Ferry. About 300 tons came from the Gosport Naval Shipyard, much of it old tools and obsolete cannon, as well as ship fittings destroyed in the fires.

To accommodate the pile of iron required for turning out the 8-inch-wide plates Brooke ordered, the doors to the Tredegar's heating furnace had to be widened. It was not until September that the first of *Virginia*'s plate began to move through the rolling mill.

During the summer and fall of 1861, John Mercer Brooke was frenetically busy. Along with overseeing the iron production, he was ordered by Mallory to design a 7-inch rifled gun for naval use. In their initial report to Mallory regarding the conversion of *Merrimack*, Brooke and the others had suggested using some of *Merrimack*'s original 9-inch

Dahlgren smoothbores, eight as broadside guns and two as pivot guns in the bow and stern. But for maximum flexibility, Brooke wanted the *Virginia's* armament to include rifled ordnance as well as smoothbores. The United States had provided the smoothbores; now Brooke set to work creating the rifled guns.

In early fall he designed the first of the Brooke rifles. It was a heavy gun, with an iron band shrunk around the breech in the manner of the successful Parrott rifles used by the North. The iron band strengthened the gun at the point of greatest stress, where the powder charge ignited in the breech, or inner end of the barrel.

Brooke wanted his gun to be able to fire either shells or "bolts," solid projectiles. The banding on the breech allowed the gun to endure the terrific stress involved in firing solid shot from a rifled barrel. On September 21, Mallory ordered the first two Brooke rifles from the already overtaxed Tredegar.

Along with the iron and guns, Brooke was called on to assist with the design of new kinds of ordnance and fuses, as well as various naval vessels. He even sat on the board that created the Confederate naval uniforms. Jefferson Davis had selected gray as the color, "universally disliked in the navy," Brooke wrote in his journal. Blue, of course, was the color of naval uniforms the world over, since in naval combat there was not the same need as in a land battle to distinguish friend from enemy by uniform color. One Confederate officer would write, "Who had ever heard of a gray sailor, no matter what nationality he served?" But gray it was, and at least the button that Brooke designed proved popular.

The original plan for *Virginia's* shield called for three layers of 1-inch iron plate, but as work progressed, Brooke began to have his doubts. In September he ordered Tredegar to roll 2-inch plate. A month before, Mallory had ordered Brooke to perform tests on the proposed makeup of the shield. Similar testing had been carried out in England, but neither Mallory nor Brooke knew the results. "We were without accurate data . . ." Mallory wrote, "and were compelled to determine the inclination of the plates and their thickness and form by actual experiment."

Because of delays in iron manufacture, due in part to Brooke's doubling the thickness of the plate, it was not until early October that Brooke could begin his tests. He decided to perform them on

Jamestown Island, far from the eyes of the Southern press, whom he considered all too zealous in reporting military secrets.

A letter written in choppy prose from Brooke to his wife, Lizzie, on October 1 illustrates just how hectic his life was at that time. "Tomorrow I shall go to Norfolk to see about the *Merrimac* returning as soon as my business is completed expect to return on Saturday. In the mean time the target will be completed and then I shall go to Jamestown Island. In this case it will be better if you can come to postpone your departure until my return from Jamestown Island. I shall speak to Cousin Virginia about whooping cough . . ."

Assisting Brooke with his tests was Lieutenant Catesby ap Roger Jones. Jones was forty years old, a twenty-five-year veteran of the U.S. navy.

Jones had served in the Far East and was a veteran of the Mexican War, though he saw little action. In 1841 he was transferred to Washington to the Depot of Charts and later the Hydrographical Office. There he worked with such future Confederate officers as Matthew Fontaine Maury and Brooke, and earned a reputation of being one of the bright scientific minds of the navy. He spent three years working with John Dahlgren on the development of the Dahlgren smoothbore. In 1856 he was given the post of ordnance officer aboard the brand-new steam frigate *Merrimack*.

By the time Jones resigned his commission, three days before the abandonment of the Gosport Naval Shipyard, he was one of the most knowledgeable ordnance experts in the U.S. Navy. Like many Virginians, such as Brooke, Jones was first attached to the Virginia Navy. In that capacity, he was one of the officers sent to help with the capture of the Gosport Naval Shipyard. After assisting in the occupation and reorganization of Gosport, Jones accepted a commission in the Confederate navy. He was assigned to command the fortifications on Jamestown Island, with the caveat that he had to build them first.

Testing on Jamestown Island

In early October the first target was prepared. It replicated the shield originally proposed for *Virginia*: 24 inches of wood covered by three layers of 1-inch iron plate and angled about thirty degrees. There was a good turnout for the first day of testing, including John Tucker, commander of the James River Squadron, William Powell, David McCorkle, and Robert D. Minor, as well as Jones and Brooke. Also there was Nelson Tift, who, with his brother Asa, was planning the construction of an ironclad in New Orleans.

The target was fired on from 300 yards using a Columbiad, a big gun developed for coastal defense, hurling an 8-inch solid shot with a charge of 10 pounds of powder. The shot crashed through the 3 inches of iron and drove itself into the wood to a depth of 5 inches.

The test was repeated seven or eight times with the same result. Clearly, 3 inches of iron was not cutting it. Brooke did not know it then, but *Virginia*'s chief antagonist would be firing 11-inch solid shot with a 15-pound charge from about 30 feet. "The 11-inch shot of the *Monitor*," he later testified, "would have penetrated the shield of the first description, I think, very readily."

The next target was constructed with two layers of 2-inch iron plate. These tests were more successful. Once again they fired an 8-inch solid shot at the mock shield, as well as 9-inch shells weighing 70 pounds. Brooke reported, "The outer plates were shattered, the inner were cracked, but the wood was not visible through the cracks in the plating." The *Virginia* would have two layers of 2-inch plate.

Unfortunately, this decision greatly increased the difficulty of production. One-inch plate was fairly easily rolled and the holes for the bolts to attach it to the wooden casemate could be punched. Two-inch plate was more difficult to roll, and it could not be punched. Apparently, the foreman at Tredegar tried, and succeeded, but then decided that punching 2 inches of iron was too hard on the machinery. As a result, every plate had to have bolt holes laboriously drilled.

To make matters worse, Porter and Brooke were designing the ship on the fly, and changing aspects of her configuration constantly. This often resulted in plates having to be redrilled, in some cases as often as four times. This increase in difficulty, along with a rise in the price of material, forced Tredegar to raise their price to seven and a half cents per pound. The delays were agonizing.

Finally, by the fall of 1861, iron plate began to roll out of Tredegar at a respectable pace. An order written on October 9 calls for the superintendent of the Petersburg railroad to furnish shipping for "2000 8 inch & other size iron bars for the steamer *Merrimack.*"

Design issues with the shield had been worked out. Tredegar was tooled up to roll 2-inch plate, and the ironworks was devoting itself almost exclusively to producing armor for the ship. But as the 8-inch-wide plates stacked up on the banks of the James River, another problem arose.

There was no way to move it the 80 miles to Norfolk.

Chapter 13

John Ericsson

The men who designed and built the CSS *Virginia*, Brooke, Porter, Williamson, and Jones, were men of talent and vision. They were at the forefront of a new technology, ahead of what anyone else in America or Europe was doing, and they managed to create a unique and powerful vessel.

But the man who designed the USS *Monitor* was a genius, a true genius, and he was years ahead of them all.

David Dixon Porter would write to Gustavus Fox and say, "Let the Government *buy* Ericsson at any price, and let him be the projector of every ironclad vessel that is to be built. There is a man with a genius not equaled in the country. Let him be our Engineer-in-chief and name his own price." Such was the acclaim that Ericsson received for his work. But he was not always viewed in such light. Far from it.

John Ericsson was born on July 31, 1803, in the village of Vångbanshyttan, in the Värmland territory in south central Sweden. He was the youngest of three, after Anna Carolina and Nils. The families of both his father, Olof, and his mother, Brita Yngström, had lived in the area for several generations.

The eastern part of Värmland where John was born is mining country, and his father was both part-owner of a mine as well as the inspector of mines at Vångbanshyttan. Olof was an educated man, schooled in Latin and Greek and mathematics. From him, John inherited his love and talent for engineering. Happily, he inherited his longevity from his mother's side: Olof died at age forty, his father at age forty-three, and his father's father at age thirty-one.

John and Nils (who would also become an engineer) were fortunate to be born to a father who could encourage and teach them the science of engineering, for which they both showed an early aptitude. Olof was a patient and careful teacher. He demonstrated mechanical principles through experimentation and took the boys along with him to survey mines and see firsthand how engineering theory was put to practical use.

To augment the boys' education, Olof sought out other professional men, in one instance giving free board to an artist in exchange for his teaching John and Nils the contemporary English style of drawing.

John's unique talents and his penchant for tireless work were clear even at an early age. As they grew older, Nils, like most young men, was drawn toward friends and socializing, while John spent most of his time in study.

In 1811, the government of Sweden revived a plan to build the Göta Canal, which would cross the country from Göteborg to Stockholm, connecting various navigable bodies of water. The idea had been bounced around for three hundred years, but at last it was to become a reality. Olof Ericsson, temporarily unemployed and impoverished, went to work on the massive construction project.

True to form, Olof was vigilant in bringing his boys to the attention of anyone who could do them good, including Count Baltzer Bogislaus von Platen, engineer-in-chief of the Göta Canal project. Impressed with what he saw, Count Platen arranged for John and Nils to be appointed cadets in the Mechanical Corps of the Swedish navy, which was overseeing the canal building, much like the modern U.S. Army Corps of Engineers. In 1815, at the age of twelve, Ericsson donned the uniform of a cadet in the Swedish navy and began making mechanical drawings for the archives of the Canal Company.

John Ericsson worked on the Göta Canal for most of his teenage years, gaining in responsibility as his abilities became apparent. By the time he was fourteen he had the direct supervision of six hundred men, though he still needed to stand on a stool to sight through his leveling instrument.

At seventeen, the ambitious Ericsson was growing restless, and looked for more in the way of military service. He resigned from the Naval Cadets and joined the 23rd Regiment Rifle Corps, an odd sort of unit that was stationed in the frozen north of the country and was employed at survey and engineering projects when not on active service.

Ericsson remained in the Rifle Corps for six years, surveying and mapping the Jämtland region where he was stationed and, in his spare time, writing and engraving a book on the engineering of the Göta Canal. As he grew tired and restless with military life, his mind turned more and more toward mechanical innovation.

Conventional wisdom was, to Ericsson, like a red cape to a bull. His was an extraordinary mind, constantly finding improvements on existing technologies and creating others that were totally unique.

At the age of twenty-two Ericsson began a quest that he would pursue all his life: to find an alternative to steam for driving an engine. Drawing on one of his father's experiments, he invented an engine that used heated air rather than steam to move a piston. This innovative "Flame Engine" proved a success on a small scale, but a poor nation such as Sweden could offer Ericsson little by way of recognition or opportunity. Ericsson's friends urged him to travel to England, which was then leading the world in mechanical development, and demonstrate it there.

Britannia

Obtaining a leave of absence from the army, Ericsson borrowed a thousand crowns to pay his traveling expenses and set himself up in England. Not quite twenty-three years old, he left Sweden in May 1826. He would never set foot in his native country again.

He arrived in London on May 18. Soon after, he arranged a demonstration of his flame engine at the Institute of Civil Engineers. Ericsson had designed his engine to be fired with good resinous Swedish pine shavings, which were not available in England. For his demonstration, he substituted sea coal, not a good choice. Burning much hotter than wood, the coal's intense output of heat destroyed the engine's cylinder, the valves, the pistons, and Ericsson's reputation as a *wunderkind*.

One witness to the demonstration was John Braithwaite, owner of an engineering firm. Despite the failure, Braithwaite liked what he saw in Ericsson and asked the young man to come work for him. It was fortunate, as Ericsson needed a job, having all but run through the thousand crowns he had borrowed. Before long, the company was renamed Braithwaite & Ericsson.

John Ericsson threw himself into his work with his usual zeal. His output was fantastic. He designed new pumping systems for mines, new boilers, a steam fire engine, and the surface condenser for the maritime steam engine. In the decade that followed his joining with Braithwaite, Ericsson patented thirty different inventions.

While still in England, Ericsson's leave of absence from the Swedish army expired, and he became, technically, a deserter. Happily, one of his admirers was the crown prince of Sweden, who not only restored Ericsson to service but gave him a captain's commission. Ericsson resigned his commission the day he received it, but continued to use the title "Captain" throughout his life. Despite the nautical implications of "captain," it was, in fact, an army rank.

During the early 1830s, Ericsson began work on an invention that could well be considered his greatest contribution to naval warfare. It had nothing to do with turrets or the iron cladding of ships. It was the screw propeller.

Paddle wheels were an inefficient means of propulsion for a number of reasons, but on men-of-war they were particularly troublesome. Set against round shot and exploding shells, paddle wheels were relatively delicate constructions. Exposed as they were, they were terribly vulnerable in battle when they could least afford to be lost.

The screw propeller was not an original Ericsson idea. By the time Ericsson began work on it, engineers in the United States, England, and elsewhere were already experimenting with screws, logging numerous failures. Indeed, while many saw the screw propeller as the obvious replacement of the paddle wheel, many others felt that it could never work. Ericsson would be the one to finally prove that it did. And while he was not the only one to patent a screw propeller, he was the first to do so, and the first to put the screw propeller to practical use.

Writing years later to a friend, Ericsson outlined the sequence of events: "1835. Designed a rotary propeller to be actuated by steam power . . ." That first propeller consisted of a wide metal hoop, to the outside of which were riveted six blades that were "segments of a screw." The blades were continued inside the hoop to where they met at a hub in the center. The whole thing looked somewhat more like the blades of a turbine than a modern ship's propeller.

Though it was just an early version of what was to come, the propeller worked. To demonstrate, Ericsson built a two-foot, propeller-driven model steamboat and set it steaming around one of the public baths in London. Small though it was, the boat made a good three knots.

As successful as his tests on the screw propeller had been, and as important a development as it was, Ericsson's innovation was met with an impressive degree of indifference. In order to show the propeller in actual use, Ericsson built the *Francis B. Ogden*, a forty-five-foot screw steamer named for the American consul in Liverpool who helped Ericsson finance the venture.

The *Ogden* was a great success. She could make nine knots through the water, much faster than a comparably-sized paddle wheeler could hope to do. Ericsson towed a 140-ton schooner at six knots and an American packet ship at four and a half knots against the tide. The demonstration astounded the men who worked the river. To the little steamer that could move with such power and no visible means of propulsion they gave the name *Flying Devil*. But the engineering community, and the Lords of the Admiralty, remained unimpressed.

Ericsson arranged for another demonstration specifically for the navy men. The *Ogden* was made fast to the admiralty barge, on board of which were the First Lord of the British navy, the Surveyor of the Navy, the famed hydrographer Captain Beaufort, and other men of influence in naval matters. Ericsson spread out on a table all the plans and specifications concerning his propeller, in which the naval men took not the least interest.

The *Ogden* got under way, steaming down the Thames and back at an easy nine knots, towing the admiralty barge "on the hip." It was a very persuasive demonstration, but the navy men were still unmoved. They thanked Ericsson, hoped he had not spent too much money on the experiment, and left.

Ericsson was astounded, though he continued to believe that the admiralty would see reason. That hope was dashed when, a few days later, he received a letter written by Captain Beaufort, explaining that the navy men had been very disappointed with the experiment.

Ericsson would later learn that the Lords of the Admiralty were already convinced, even before seeing the *Ogden* in action, that a ship would not steer if its propulsion was at the very stern of the vessel, as was the case with the propeller. How they thought Ericsson managed to get on and off the dock and up and down the river without steering they did not explain. Ericsson's breakthrough propeller, as his biographer William Church wrote, "confused the mind of the average Englishman, who hates a thing merely because it is new."

For all of the innovations coming off of John Ericsson's drawing board, few were making any money. Braithwaite & Ericsson spent much of its existence teetering on the brink of insolvency, and the *Ogden* pushed it over the edge. By the summer of 1837, the company was £12,000 in debt, and Ericsson owed a huge sum to his tailor, which he could not hope to pay. Unemployed, ignored by the Lords of the Admiralty, broke, John Ericsson was now unceremoniously tossed into the debtor's prison in Fleet Street.

It is uncertain how long Ericsson remained in prison, but it was probably not long. In the face of generally hard economic times, Parliament had passed an act for "the relief of insolvent debtors." Ericsson took advantage of that to secure his release.

He had just about had his fill of England.

And that was fortunate, because soon after his release, Ericsson would meet Lieutenant Robert F. Stockton, U.S. Navy, and Stockton would completely alter the trajectory of Ericsson's life.

Chapter 14

The *Princeton*

Lieutenant Robert F. Stockton was the oldest son of an old and powerful New Jersey family and grandson of a signer of the Declaration of Independence. He was tall and handsome, a daring naval officer, immensely wealthy, a smooth orator, and perfectly aware of all of his own attributes. He was also a man of vision, willing to take risks (it was that quality that nearly ruined his naval career) and ready to invest where he saw opportunity.

In 1837, Stockton was in London looking for investors for a project into which he himself had sunk a fortune: the Delaware and Raritan Canal. It is most likely that fellow New Jersey native Francis Ogden introduced Stockton to Ericsson. In short order, Ericsson had the American naval officer aboard the steamer *Francis B. Ogden*, steaming down the Thames at her impressive rate of speed, her direct-acting engine turning the submerged propeller and, the Lords of the Admiralty's assurances notwithstanding, her rudder steering her quite well.

Stockton did not suffer from the narrow worldview that afflicted the British naval officers. He saw immediately the significance of what Ericsson had created. He told Ericsson, "I do not want the opinions of your scientific men; what I have seen this day satisfies me."

The propeller-driven ship seemed to Stockton just the thing for towing barges in his canal. He commissioned Ericsson to build two vessels on the *Ogden* model but twice the size. To Ericsson, who was pretty near rock bottom, Stockton's enthusiasm, his trust, and his money must have seemed like manna from heaven.

Stockton returned to America and was promoted to captain in the navy. On his way to join the Mediterranean fleet, he stopped in London with dispatches to the Court of St. James. He was in time to witness the trial of the first of his Ericsson boats, which he modestly named the *Robert F. Stockton.*

The *Stockton* was built of iron by John and Macgregor Laird on the Mersey, the same company that would later build the famous Confederate commerce raider *Alabama*. She was fitted with a double-cylinder, direct-acting engine and Ericsson's patented screw propeller. On January 12, 1838, she was tried on the Thames in front of Stockton, Francis Ogden, and various other dignitaries invited to witness the event. She performed beautifully. It was the beginning of the end for the paddle wheel.

The *Stockton* carried a schooner rig, and under canvas she set sail for New York, arriving in late May. Despite a rough passage, she suffered "no serious disaster except the loss of one seaman, who was washed off this little cockle-shell by one of the seas which were constantly sweeping her decks." The ship caused quite a stir in New York City, not because of her unique screw propeller and direct-acting engine, but because people were astounded that so small a vessel had made a transatlantic voyage.

The *Stockton*'s sailing rig was taken down and the ship was sent off to the Delaware River, where she spent the next twenty-five years working as a successful river and canal tug. But her namesake, Captain Robert Stockton, was not done promoting Ericsson and his propeller. He had only just begun.

In March 1839, ten months after the arrival of the *Ogden* in New York City, Congress passed an authorization for building three men-of-war. During the preceding year, Stockton had been lobbying hard for steam power and for Ericsson's screw propeller. His energy and influence carried the day. Stockton was assured that one of the ships would be built to the design he advocated. Still in England, Ericsson began designing the vessel he and Stockton had envisioned.

After his release from prison, Ericsson had earned his living as superintending engineer of the Eastern Counties Railroad. In November 1839 he quit that position to come to America, motivated by Stockton's assurances that the navy contract would come through and that he, Stockton, would make Ericsson wealthy and famous. He set sail for America. It was a rough crossing, and Captain Ericsson was terribly seasick.

Arrival in America

Unfortunately, Stockton was running into the kind of close-mindedness among senior naval officers that Ericsson had experienced in England. "On arriving here," Ericsson wrote five years after, "I soon found that Captain S. had not that power with the administration as he had told me in England—where he once assured me he could get my propellor introduced in the American navy at once." It would be more than a year and a half after his arrival in New York before Ericsson would see his "big frigate" begun.

Showing his usual disregard for fiscal management, Ericsson took up residence in the famous Astor House hotel and began to market his inventions on his own. Not two months after his arrival, he won a gold medal from the Mechanics' Institute of New York for his steam fire engine design. He also designed a series of small propeller-driven barges for various commercial interests. Four of them were for Stockton, for which Ericsson was never paid.

During this time, Ericsson established a relationship with the junior partner of the Phoenix Foundry, an ironworks on the Hudson River. That partner, Cornelius Delamater, would become a lifelong friend and business associate. Two decades later, Delamater would send Cornelius Bushnell to Ericsson's door.

Finally, by the fall of 1841, things began to happen regarding the construction of the "big frigate." On September 21, Stockton received clearance to begin work on a "steamer of six hundred tons on the plan proposed by Captain Stockton; steam to be the main propelling power upon Ericsson's plan."

Stockton summoned Ericsson to his home in Princeton, New Jersey, and there Ericsson came up with a rough sketch for a 600-ton, propeller driven naval vessel. It was considerably smaller than the 2,000-ton vessel Ericsson had designed in London, but it was at least an opportunity for Ericsson to put a number of his innovations into practice.

Ericsson returned to New York and began to design in earnest. He did it in the Ericsson style, creating not just the engine and propeller but the ship itself, as well as the guns and the gun mounts. As would be the case with *Monitor*, every inch of the vessel carried the Ericsson stamp.

Prior to their meeting, Stockton had written to Ericsson asking Ericsson to let him know what commission he would deem proper for the use of his patents. Stockton suggested that Ericsson should only be paid after the ship was put to the test.

As dubious an offer as this was, Ericsson made his situation even worse by replying, "I shall be satisfied with whatever sum you may please to recommend, or the Government see fit to pay for the patent rights." Ericsson seemed to feel in this case, as he did in many others, that the sheer genius of what he had created ensured his future monetary reward. Whatever his reasons—excessive trust in his fellow man, arrogance regarding the obvious value of his work—Ericsson was no Thomas Edison when it came to squeezing profit from his inventions.

It took nearly two years to design and build the ship that would be named *Princeton*, after Stockton's hometown. On September 7, 1843, she was launched with considerable fanfare and publicity, arranged by Robert Stockton largely to promote Robert Stockton. At a lavish party given by Stockton on the day of the launch, the navy captain told his guests that he had "been all over the world in search of a man who could invent or carry out what he thought was necessary to make a complete ship of war." Then he introduced John Ericsson as that man.

It was no accident that this lukewarm accolade left the listener with the clear idea that the ship was the brainchild of Robert Stockton, and that Ericsson had merely been an instrument in carrying out Stockton's grand scheme. As time went on, Stockton would become more egregious in pushing Ericsson's contributions to the background. That was, until the disaster.

USS *Princeton*

The *Princeton* was enormously successful, and her list of firsts is staggering. She was the first iron, propeller-driven man-of-war, the first to
employ a direct-acting engine, and the first to have her machinery and
boilers below the waterline, safe from enemy shot. She was the first to
burn smokeless anthracite coal so she would not give her presence
away by a great plume of black smoke. She was the first to employ a
collapsible smokestack and the first to bring a forced draft of air to the
fires in the boilers through the use of centrifugal blowers rather than
natural draft.

All this and more Stockton pointed out in a lengthy report he wrote
to the secretary of the navy on February 5, 1844. Not once was Ericsson mentioned. He had not yet even been paid for his efforts.

With the *Princeton* complete, Stockton began a publicity tour,
showing "his" ship off to the fascinated crowds who jammed water-
fronts to see her. He took her to New York to receive her guns and engaged in a speed trial with the *Great Western*, one of the fastest
steamers of the day. Meeting the outbound *Great Western* on the East
River, *Princeton* steamed up to her, passed her, circled her twice, then
returned to the dock. It was an astounding feat, and more astounding to
the crowds watching from shore, as *Princeton* had no visible means of
propulsion: no paddle wheels, no sail set, and, thanks to the anthracite
coal, no smoke coming from her smokestack.

Stockton never claimed to have designed or built the ship, but he did
little to correct the impression, shared by many, that he had. When it
came time to sail from New York back to Washington, Ericsson was invited to come along and help Stockton demonstrate the vessel for members of Congress and the cabinet. A meeting was arranged. Ericsson
would wait at the foot of Wall Street and the *Princeton*, after getting un-
derway, would send a boat for him.

Ericsson made the rendezvous, standing in the cold of a New York
City February. He watched as his ship, the culmination of so much of
his engineering invention, steamed downriver. He watched her steam
past. He watched as she steamed through the Verrazano Narrows, leav-

ing him standing there on the dock, forgotten. As hurt and angry as Ericsson might have been at that moment, he could not have realized how lucky he was to be left behind.

When Ericsson first came to the United States he brought with him, along with various plans, models, and his fabulous wardrobe, a cannon. Ignoring the old admonition to travel light, Ericsson packed a huge wrought-iron gun, 15 feet long, capable of firing a 12-inch, 212-pound shot. This gun he placed aboard the *Princeton*. It was named *Oregon*, after the territory over which the United States and Great Britain were currently disputing ownership. Relations between the two countries had never been warm, and much of what the U.S. military did, and the navy in particular, was with an eye toward fighting the British again, or beating them in an arms race.

After landing his gun in New York, Ericsson began systematically testing it. Proof firing created cracks in the barrel just behind the trunnions. This problem Ericsson corrected by shrinking $3\frac{1}{2}$-inch wrought-iron hoops around the barrel, a strengthening technique that would become nearly universal in Civil War era guns. Ericsson tested the reinforced gun in a hundred test firings, and found it completely sound.

Stockton wanted to have a gun, too. Borrowing Ericsson's plans for *Oregon*, he had another cast at the Hamersley Forge in Philadelphia and then sent to the Phoenix Foundry in New York to be bored under Ericsson's supervision.

Stockton's gun was made on the *Oregon* model, but was heavier throughout, the breech being a foot greater in diameter. It weighed 10 tons, the largest piece of forged iron in the world. Because of the heavy breech, Stockton considered Ericsson's shrunken bands unnecessary. Stockton dubbed his gun *Peacemaker*.

On Wednesday, February 28, Stockton welcomed aboard the *Princeton* the cream of Washington, D.C.'s elite: President John Tyler and his party, which consisted of most of his cabinet and their families, as well as influential civilian and military personnel. A little over a week before, Stockton had hosted members of the House and Senate, but this day was the crowning moment of his promotional efforts for his ship and himself.

With the guests aboard, *Princeton* weighed anchor and set off down the Potomac River. It was a lovely day, and Stockton was eloquent in his praise of the ship and her qualities. They demonstrated the ship's guns at a broad place in the river, to the delight of the hundreds of men and women onboard. They sat down to a sumptuous meal, where President Tyler led the toast, "I give you the *Oregon*, the *Peacemaker*, and Captain Stockton!"

As the afternoon wore on, the *Princeton* turned upriver again and steamed back toward Washington. While some of the guests continued to tour the ship, Stockton and many others retired below for a continued round of toasts. As Stockton rose to give a toast, an officer entered the cabin and told him that some of the guests requested that the big guns be fired again.

"No more guns tonight," Stockton said, and returned to his toast. Soon the officer was back, informing Stockton that the secretary of the navy himself wished to see the gun fired.

A request from the secretary of the navy was tantamount to an order in Stockton's mind, so he proceeded on deck to arrange the demonstration. Ericsson had earlier recommended using the *Oregon* for demonstrations, most likely out of simple prejudice for his own gun, as he seemed to have no early misgivings about the *Peacemaker*. In any event, Stockton chose his own gun, and 50 pounds of powder was rammed down the huge barrel.

Those guests who were interested gathered around. President Tyler had originally been one of them, but he wandered off as the gun was made ready. The secretary of war, Wilkins, also begged off, saying, "I may be Secretary of War, but I have no heart for this firing."

With the powder rammed home and the guests standing clear of the recoil, Stockton ordered the gun crew to fire. With a jerk of the lanyard, the powder in the barrel ignited and the *Peacemaker* blew apart.

Chapter 15

"Another Ericsson Failure"

The blast of the exploding gun hurled Stockton to the deck. He lay there, stunned, burned on his hands and face, unsure of what had happened.

And then through the fog in his brain and the great cloud of smoke on the deck came the screams of his guests, the cries of many, many wounded men, and Stockton knew.

He pulled himself to his feet, and took command of the situation with the authority that had made him a successful naval officer. He ordered mattresses and blankets brought up, ordered the women below to be informed of what had happened, but to be kept there. He ordered the American flag to be draped over the bodies of those killed.

Six men lay dead. The secretary of state, Abel Upshur, had been killed instantly, torn apart by the flying metal, as had the secretary of the navy, Thomas Gilmer, who had given the tacit order to fire the gun. His wife, uninjured, stood just a few feet from him. Also killed were Captain Beverly Kennon of the navy, two civilian VIPs, and President Tyler's longtime servant. Seventeen seamen were seriously wounded, though none died. Incredibly, no women were injured or killed.

The accident was a national tragedy. Flags were flown at half-mast. Funeral services were held at the White House. Congress adjourned.

But soon they were back, and they wanted answers. And suddenly Stockton felt the need to have his hitherto forgotten partner Ericsson by his side, undoubtedly hoping the engineer would take some of the verbal shrapnel he knew would be flying.

The *Princeton*'s executive officer wrote to Ericsson, requesting his presence in Washington. Ericsson declined, acidly pointing out that since he had not been aboard the ship he could offer no opinion on the way the gun was loaded and fired. "How differently should I have regarded an *invitation* from Captain Stockton a week ago!" he wrote two days after the accident. "I might then have had it in my power to render good service and valuable counsel. *Now* I can be of no use."

Ericsson went on to point out that since Stockton had in his possession the plans of the *Peacemaker*, there was nothing more that he could add in that regard, and any intelligence to be gleaned from the forge that cast the gun or the Phoenix Foundry that bored it, he could just as easily get while remaining in New York.

Stockton considered Ericsson's refusal to come to Washington a betrayal, and he was not one to suffer a betrayal without retaliating.

Two weeks later, Ericsson submitted to the secretary of the navy his bill for the work he did on *Princeton*, including a five-dollar *per diem* for 230 days' work and other charges for his inventions. The total was for $15,080.

The navy department referred the bill to Stockton. Two months later, after constant urging from Ericsson, the navy finally sent a reply. Stockton would not approve the bill. He said it was in direct violation of their agreement.

Nine days later, Stockton sent a long and astoundingly disingenuous report to the secretary of the navy detailing his version of Ericsson's participation in the *Princeton* project. After going on at some length about the insignificance of the relationship he had had with Ericsson prior to the *Princeton*, Stockton wrote, "Captain Ericsson came to the United States without my invitation or approbation, and allow me to further add, much to my surprise and annoyance. Having thus thrust himself on me, and believing him to be a mechanic of some skill, *I did not employ him, but I permitted him, as a particular act of favor and kindness*, to superintend the construction of the machinery of the *Princeton* . . ."

Such had become the man for whom Stockton had "been all over the world in search of."

The correspondence, appeals, and recriminations went on for eleven long years. Finally, in 1855, Congress passed an act that established a Court of Claims to settle disputes outstanding between the government and individuals, reserving for itself the right to approve or disapprove the court's findings. The following year the Senate ordered Ericsson's papers to be delivered to the court for consideration. The court found in Ericsson's favor and recommended that $13,930 be paid him.

That recommendation went to the Senate for approval. The senator from Florida, chairman of the Naval Committee, Stephen R. Mallory, rose in defense of the payment. "The *Princeton* is the foundation of our present steam marine," he said. "It is the foundation of the steam marine of the whole world . . ."

The matter was referred to a committee, and there it went into legislative suspended animation. Nothing was ever done, no money was ever appropriated. All the work Ericsson had done on *Princeton*, and all the effort he had undertaken to secure payment for that work, was for nothing.

Bushnell the Promoter

And so the John Ericsson who showed Cornelius Bushnell his idea for a shotproof, turreted iron battery was not a man much inclined to do business with the U.S. Navy. In the years since the *Princeton* disaster he had made quite a bit of money and spent quite a bit as well, much of it on developing his inventions. In the end, his concept of a "caloric engine," which used heated air rather than steam, had proved a great success when used in smaller engines. Thousands of the calorics had been sold, making Ericsson, if not wealthy, then at least comfortable.

In all likelihood, Bushnell knew all of Ericsson's history, even if only in a general sense. He apparently made no effort to persuade Ericsson to present his ideas to the board, but rather asked if he might take the model and the plans and present them himself. Bushnell the well-connected lobbyist knew that he, and not the hot-tempered, once-bitten Ericsson, would stand a better chance in front of the ironclad board.

Ericsson let Bushnell have it all, with his blessings. He allowed Bushnell, whom he had known less than twenty-four hours, to make whatever business arrangements he wished. By way of payment, Ericsson said he would accept whatever "interest or compensation" that Bushnell thought was fair. Either Ericsson genuinely did not care about monetary reward for his battery, as he wrote to Lincoln, or he was showing his usual poor judgment in business matters. Fortunately for him, Bushnell was no Robert Stockton.

Cornelius Bushnell left New York that evening with the model and the plans and took the train to Hartford. He knew that Gideon Welles, still in the process of moving to Washington, had gone there to pack up his household and his family. It was the first time Welles had been home since leaving for Washington in February. Excited by this novel concept, and with time running out to submit it to the ironclad board, Bushnell felt a great sense of urgency, so much so that he passed right through New Haven—his own hometown—without stopping.

In Hartford, Bushnell showed the model and the plans to Welles, assuring the secretary that he "had found a battery which would make us masters of the situation, as far as the ocean was concerned." Those were words that Welles wanted to hear. He was impressed with the concept and swept up by Bushnell's enthusiasm. But Welles understood, as did Bushnell, that convincing the ironclad board to back so radical a notion—and worse, a radical notion designed by John Ericsson—would be no easy task.

Welles urged Bushnell to go at once to Washington and present Ericsson's plans to the board, assuring Bushnell that he would follow in a few days to help with the proposal. Bushnell left the next day, a Saturday, and arrived back in Washington on Sunday morning. Anticipating the battle he would be waging with the ironclad board, Bushnell decided to line up some heavy artillery.

After breakfast that morning, he hired a carriage and went around to see his friends and business partners, John A. Griswold and John Winslow, both ironmakers from Troy, New York, who happened to be in

Washington at the time. The two men were very well connected politically, even more so than Bushnell. Griswold in particular was active in politics, and two years later would be elected to Congress. Bushnell had already entered into an agreement with them to provide iron plate for *Galena* if that contract was approved. Now he had a new proposition.

Bushnell took Griswold and Winslow on a quiet ride through the suburbs of Washington and laid before them Ericsson's plans. They, too, were able to see the vessel's merits (as well as the amount of iron plate that would be required to build her). Ericsson had given Bushnell leave to negotiate whatever terms he wished. Now he asked Griswold and Winslow to come aboard as partners, taking a half-interest in the vessel while Bushnell retained the other half for Ericsson and himself.

Bushnell went on to explain to his associates why the battery would be a hard sell to the ironclad board. As it happened, and as Bushnell undoubtedly knew, both the Troy men were good friends with fellow New Yorker and current secretary of state William Seward. All three men felt they should seek Seward's endorsement, even though they already had the unqualified support of the secretary of the navy. Such was the influence that Seward wielded in the Lincoln administration.

The next morning, Monday, the three men went to see the secretary of state. Seward was impressed with Ericsson's ship and wrote them a "strong letter of introduction" to President Lincoln.

That same afternoon they called on the president, who was also impressed with what he saw. Lincoln readily admitted to knowing nothing about watercraft, save for flatboats, of which he claimed to have once been a master. "And as the little boat or model we showed him the plan of was flat as need be," Bushnell said, "he understood the good points from the start."

Lincoln had no authority over the ironclad board's decision, and told Bushnell and company as much, but he so liked the idea of the floating battery that he offered to accompany them during their presentation.

Back to the Ironclad Board

The following day at 11:00 A.M., Bushnell made his pitch. True to his word, Lincoln came to the meeting, sitting silent as the plans were discussed, lending the weight of his office, if not his words, to Bushnell's presentation. Two of the three board members were there—Davis was not present—along with a number of naval officers on an unofficial basis as well as Assistant Secretary of the Navy Gustavus Fox.

Bushnell presented the plans with enthusiasm, but they were greeted with quite a different reaction. The novelty of the idea took the navy men by surprise. Most of those present ridiculed it and thought it was absurd. A few, however, most notably Admiral Smith and Commodore Paulding, were intrigued and thought the plan worth pursuing.

As the meeting drew to a close, Smith turned to the president, who had remained silent throughout, and asked his opinion. Lincoln stood up to his impressive height and drawled, "All I have to say is what the girl said when she stuck her foot into the stocking. It strikes me there's something in it."

Fortunately for Bushnell, the two naval officers who actually supported the Ericsson plan, Smith and Paulding, were the two who counted. But with an idea so unique, Smith and Paulding did not want to give a thumbs-up without Davis's approval. The next day Bushnell appeared before the entire board and once again gave his pitch. He was eloquent and persuasive. He took careful note of each man's reaction, and went away satisfied that he had convinced them to approve the project.

He was wrong. The following morning he presented himself before the board only to discover that, in the hours since his last meeting, there had been sufficient negative talk to sour the men on the idea. The board was afraid that they were about to "father another Ericsson failure."

Bushnell pointed out that there had never been an Ericsson failure, that the *Princeton* was a triumph and that the explosion of the *Peacemaker* had nothing to do with him. Ericsson's caloric engine, even if it had not been made to work well on a large scale, was a great success in the small engines he was manufacturing.

At last, through his most eloquent persuasions, Bushnell got Smith and Paulding to agree to approve the project if Davis would as well. But Davis would not budge. He told Bushnell to "take the little thing home and worship it, as it would not be idolatry, because it was made in the image of nothing in the heaven above or on the earth below or in the waters under the earth."

Here was the culmination of a week of manic activity that had seen Bushnell careening between Washington, New York, Hartford, and back, pitching the idea to the secretary of the navy, Winslow and Griswold, the secretary of state, the president, and several times to the ironclad board.

Dejected and spent, Bushnell reported the board's final answer to Welles, who had since returned to Washington. But Welles was not ready to admit defeat. He suggested that there was still a way to influence Davis, and Bushnell caught his meaning: get Ericsson to come to Washington and present the proposal himself.

A Final Effort

Bushnell recognized that this was their last chance. Ericsson was a dynamic speaker, a great presence, a "full electric battery in himself." Bushnell later wrote, "I never met a man possessed of more power to magnetize and carry his audience with him than Captain Ericsson." Unfortunately, Ericsson had sworn an oath to himself that he would never set foot in Washington again. Only a man of Bushnell's persuasive powers could hope to lure him back.

Bushnell once again boarded a train and headed north to New York. Rattling along in the train car, he tried to devise a means by which he could persuade Ericsson to forget his oath. He recalled the pride with which Ericsson had shown him the gold medal from Napoleon III. And then it came to him. "I will get him on his vanity," he thought, and he knew just how to do it.

He arrived at Ericsson's at nine o'clock the next morning. The passionate inventor belied his lack of interest when he, rather than his house girl, Ann, met Bushnell at the door and, by way of greeting, demanded, "Well! How is it?"

Bushnell had his script worked out in advance. "Glorious," he replied.

"Go on, go on," Ericsson said. "What did they say?"

"Admiral Smith says it is worthy of the genius of an Ericsson."

Bushnell could see the pride in Ericsson's eyes. Ericsson was so convinced of the brilliance of his invention, and so certain that the rest of the world must see it as well, that it did not occur to him that this was all very unlikely.

"But Paulding, what did he say?" Ericsson asked.

"He said, 'It's *just* the thing to clear the "Rebs" out of Charleston with.'"

Bushnell gave a slight but theatrical hesitation. Seeing that, Ericsson asked, "How about Davis?"

"Oh, Davis!" Bushnell said, "he wanted two or three explanations in detail which I couldn't give him and so Secretary Welles proposed that I should come and get you to come to Washington and explain those few points to the entire board in his room tomorrow."

Bushnell had hit the right note, and appealed to the right weakness in Ericsson's character. Convinced that the ironclad board saw the genius of the invention and wanted only to ask a few friendly questions, Ericsson entirely forgot his promise to never go to Washington again. In fact, he was absolutely eager. "Well, I'll go," he told Bushnell. "I'll go tonight."

Bushnell had succeeded admirably in getting the inventor to travel to Washington. There was now just one remaining problem: how would Ericsson react when he discovered what the ironclad board really thought of his invention? Bushnell, for one, did not care to be there to witness it. "I had remained in New York," he wrote, "not just fancying the presence of Captain Ericsson when he should first meet Captain Davis . . ."

Welles was also anticipating this problem. He approached Admiral Smith and requested that Ericsson be "treated tenderly, and opportunity given him for full and deliberate hearing."

Ericsson arrived the next day and presented himself to the board. He walked into the room thinking that he was facing three enthusiastic supporters with a few questions, but soon realized that that was not the case. "I was very coldly received, and learned to my surprise that said Board had actually rejected my *Monitor* plan," he wrote.

Even a man of reasonable temper might have been put out at that point, and Ericsson was not always a reasonable man, particularly after having spent all night on the train from New York City. Worse, he had been set up for another humiliation by the U.S. navy. Furious, Ericsson's first reaction was to storm out of the room, but he stopped himself long enough to ask the board why they had rejected his proposal.

Admiral Smith told him that the board felt the vessel lacked stability. Here, once again, was the old concern that had sunk many of the other proposals and had almost sunk *Galena*, though one wonders how the board could have thought a vessel that was as much a raft as a ship might have been unstable. The board's concerns are indicative of the revolutionary and untried nature of iron plating on ships. Those men raised in the wooden sailing navy simply did not know what would happen.

The stability issue was another red cape waved in Ericsson's face, and he would not let it go unchallenged. He launched into an elaborate explanation of why there were no stability problems with the ship. The more he talked, the more his blood boiled. He ended his presentation with a typical Ericsson admonition: "Gentlemen, after what I have said, I deem it your duty to the country to give me an order to build the vessel before I leave the room."

Smith, Paulding, and Davis did not interpret their duty in quite the same way as Ericsson, but his demonstration had impressed them enough that they were willing to reconsider. There were a couple of things about the battery that they found very attractive. One was the price, $275,000, which was about $40,000 more than *Galena* but almost a third the price of the ship that would become *New Ironsides*. The other appealing thing was the time required to build the ship, which was only one hundred days. Welles and the ironclad board were very aware of the work already taking place in Norfolk, and Confederate plans for ironclads in New Orleans and Mobile. No other vessel proposed to the board could get into action that swiftly.

After talking amongst themselves for a few moments, the board asked Ericsson to come back that afternoon, which Ericsson agreed to do.

At 1:00 P.M. Ericsson returned, and his greeting was more cordial this time. Paulding asked to meet with Ericsson privately, and when the two men were alone, Paulding asked for a further explanation of the stability factor. Ericsson once again explained, this time aided by a diagram he had drawn up in the hours since the first meeting. Paulding

was very impressed. He told Ericsson, "Sir, I have learned more about
the stability of a vessel from what you have said than I ever knew before." That comment speaks to Ericsson's ability to explain complicated
theories, as well as to Paulding, and no doubt the other officers' lack of
understanding regarding that issue.

Once again Ericsson was asked to return later in the day. When he
did, he was ushered into Secretary Welles's room. Welles had polled
the ironclad board and found that they were ready to support the Ericsson floating battery. The ship would be built.

Welles asked Ericsson to begin work immediately, even though
there was no formal contract, and Ericsson agreed. He returned to New
York and there turned his enormous energy toward the creation of a
ship that was to be the crowning achievement of his life.

So quickly did Ericsson mobilize production that the keel plates of
the ship had already passed through the rollers at the iron mill before
the contract was written. And that was a fortunate thing. Because if Ericsson had known what would be called for in the contract, then, as he
himself wrote, "The *Monitor* would never have been built."

Chapter 16

"She Will Be Your Coffin"

On the day that Abraham Lincoln was elected president, twenty-five year old First Assistant Engineer H. Ashton Ramsay of the USS *Niagara* was half a world away. Like many officers in the U.S. navy, those events that were raising such passions at home seemed remote and somehow unreal. "We were so far from home," Ramsay wrote, "that many could not give serious thoughts to these matters."

The United States, wrapped in autumn cold and feeling the first stirrings of war, must have seemed far away indeed as the *Niagara* plowed her way through the Indian Ocean, bound for the Middle East from Japan. Under the clear sky and the hot equatorial sun, the ship's officers dropped mock ballots into a cigar box that was doing duty as their ballot box.

The outcome of the *Niagara*'s election was quite different from the real one. The ship's officers elected the Constitutional Union party's candidates, John Bell, a slave owner from Tennessee, and Edward Everett of Massachusetts, president and vice president respectively.

The Constitutional Unionists were the most conservative of the parties running, and their platform specifically took no stand on any of the issues that were currently threatening the Union. Coming in second was John Breckinridge of Kentucky, the Southern Democratic candidate. Presumably Lincoln and Douglas received some votes, but not enough for Ramsay to mention.

Ramsay, a native of Virginia, felt the *Niagara*'s election result was an illustration of the conservative nature of naval officers, since the *Niagara*'s officers were, by his report, from all regions of the United States. It might also be an indication of how out of touch with national politics these men on foreign station were. Back home, Bell won only Virginia, Kentucky, and his native Tennessee. Breckinridge won the South. The man who barely registered in the *Niagara*'s polling became president.

When the *Niagara* arrived at last in "Arabia," the crew learned of Lincoln's election. The American newspapers they read were filled with stories of secession and disunion. Half a world away, the sailors did not believe that it could ever come to that.

The emotions that Ramsay and his shipmates experienced in that utterly foreign place are perhaps one reason that Southern naval men were so torn in their decision to go South. To them, issues of Democrat versus Republican, Free Soilers and Pro-slavery parties, were abstractions. The constant in their lives, the thing that held them together when they were on the other side of the globe, was their devotion to the navy and the United States of America.

"We had gazed on the flag with emotions of pride on every sea," Ramsay wrote, "never entering a port, however unfrequented or obscure, without being greeted by the sight of the American flag flying from the peaks of one of our merchantmen." Their reality was the U.S. navy and the flag it defended. Unlike army officers, who tended to be stationed in one state or another, the navy men, particularly those on foreign station, felt they represented the nation as a whole. For that reason, the moral dilemma seemed greater to the navy men than to those in the army, and proportionally fewer of them went South.

Mary Boykin Chesnut understood this conflict when she wrote in August 1861, "Then the navy men came under discussion. One said, 'They are like the people St. Paul cited who were sawn asunder.' There is an awful pull in their divided hearts. Faith in the U.S. Navy was their creed and their religion. And now they must fight it—and worse than all, wish it ill luck."

Many Southerners, including Chesnut, felt that Southern naval officers should have sailed their ships into Southern ports and turned them over to the Confederacy, reasoning that they belonged as much to the Southern states as the Northern. But not one officer did so.

Niagara sailed for home, stopping at Cape Town before crossing the Atlantic. It was April 19, 1861, when they made landfall in Massachusetts. The seas were running high and the rain falling hard as the frigate stood into Massachusetts Bay and hove to for a pilot. But even more than pilotage, the men of the *Niagara* wanted news. Standing in the driving rain, crowding around the pilot, they demanded the latest word.

"The Union is smashed and gone to hell," the pilot informed them. He handed them a bundle of Boston and New York newspapers. "See for yourself."

Ramsay, who had been anticipating his promotion to chief engineer, was now facing the biggest decision of his life, and he would have only a few hours to make it. *Niagara* steamed for the navy yard but was ordered instead to anchor off of Boston, quarantined for possible virulent ideology. A loyalty oath, which all officers were expected to sign, was brought aboard.

The captain assembled the men in the great cabin. "The question of signing the oath or not, each man must decide for himself," he explained. There were tears in his eyes. He told them that the decision had to be made immediately, as the *Niagara* was to sail for Hampton Roads and blockade duty that same day.

Ramsay and eleven of his fellow officers chose to go South and they were put ashore, a bad time for Southern sympathizers to be in Boston. The next day the papers were full of reports of the 6th Massachusetts being fired on while marching through Baltimore. The same papers published the names of the twelve men from *Niagara* who had opted to leave.

An ugly mob assailed the Southerners as they were trying to leave the city. Finally the Southerners were arrested, on what charge Ramsay does not say, perhaps for inciting a mob to riot. Happily, Flag Officer Stringham, in command of the naval yard, took charge of the prisoners and put them safe onboard a navy receiving ship. There they remained for ten days, treated well by their sympathetic fellow officers, until word came from Washington to release them.

Ramsay, dressed as a civilian (which, technically, he was), made his way through hostile country to Charles County, Maryland. It was not an easy trip, and the last part, crossing the Potomac River under the guns of the Union blockading fleet, was accomplished at night in an open boat. But finally, H. Ashton Ramsay was home.

Unique Qualifications

In June, Ramsay traveled to Richmond, where he found the city caught in the chaos of seating the Confederate government, "all bustle and confusion." As an engineer and a U.S. navy veteran, Ramsay understood better than more sanguine Confederates what his new country was up against. "Knowing well the vast resources of the north, in everything that goes to make up the sinews of war, I did not see much to reassure me."

Thinking there was nothing happening in the naval line, Ramsay joined the army. He was actually in camp, beginning training, when he was summoned to the navy department. Ramsay had a few qualifications that Mallory was looking for. The Confederate navy was desperate for qualified engineers. But more to the point, Ramsay had served as second assistant engineer aboard USS *Merrimack* during the frigate's final cruise. His boss was Chief Engineer Alban Stimers, who would play a crucial role in building the *Monitor*, and who would work the turret engine during the battle between the ironclads.

Mallory told Ramsay to report to the Gosport Naval Shipyard. The engineer working with Williamson in refurbishing the *Merrimack's* engines had been relieved for "insubordinate & disrespectful conduct," and Ramsay was ordered to assume his duties.

Ramsay understood the difficulty of this undertaking. "I was . . . familiar with the machinery and fully cognizant of all its shortcomings, which I set to work to correct as far as possible, with the limited means at my command." *Merrimack's* engine, though well designed, was poorly constructed using deficient material, rendering the entire thing "radically defective," in Ramsay's opinion. The condensing apparatus, the arrangement of the steam chests, the valve mechanisms, and the reversing gears were all unreliable. The engine was built at the West Point Foundry in Cold Spring, New York, the principal manufacturer of cannons and other large castings in the United States. The foundry's superintendent was none other than Robert Parrott, inventor of the Parrott gun. Despite the foundry's credentials, the engine was so bad that the navy had decided to condemn it. *Merrimack* was slated to have her engine pulled when she fell into Confederate hands.

As bad as it was, it was the best that the South could muster.

Ramsay arrived at the shipyard as the carpenters were clearing away the charred debris that had once been the *Merrimack's* sides. He joined Williamson and the yard's other mechanics in overhauling and refurbishing the ship's engine, which had not improved after nearly a month and a half underwater.

Merrimack's engine was a massive affair, enormous for its time. Its sheer size pushed the design envelope, considering the materials available which may have added to its notorious unreliability. Horizontal back-acting with twin cylinders, its drive came from two pistons, each six feet in diameter, with a three-foot stroke. Like all reciprocating steam engines, *Merrimack's* required a great deal of maintenance and adjustment, and it is also likely that failures in those areas were partially to blame for the engine's infamous reputation.

As the work on the machinery progressed, the casemate rose up over the hull. The old hull had been cut down to the level of the berth deck, the *Merrimack's* lowest deck and the only one to survive the fire. The length of this deck was around 258 feet. The *Virginia's* length overall, including the cast-iron ram and the fantail, would be around 278 feet.

Virginia's foredeck, made up of the forward end of the berth deck not covered by the shield, was around 28 feet long. The deck was 2 feet below the intended waterline, and would have been submerged, except that it was enclosed by a 4-foot-high bulwark made of wood. The bulwark— part of Brooke's concept—prevented water from piling up on the front of the shield and gave the crew a dry place to work the anchor gear. Both the foredeck and the afterdeck were covered in 1-inch-thick iron plate.

The hull areas at the very bow and stern were heavily built up with timber to form almost solid ends. Behind this timber blocking, collision bulkheads were installed, allowing *Virginia* to safely smash herself into another vessel. According to Brooke, "We always intended that she should be a ram. All ironclad vessels are built as rams."

To protect the vulnerable rudder and propeller, Porter placed a heavy fantail on the stern. This was in essence an extension of the deck that overhung the rudder and propeller, built up out of 1-foot-square timber with no iron on top. With the ship floating on her intended waterline, this fantail would be 2 feet below water, protecting the propeller and rudder from plunging gunfire. Any vessel attempting to ram the after end of the *Virginia* would hit the fantail first and with luck be deflected, though this area remained terribly vulnerable.

The shield itself was around 150 feet long. Inside the casemate, just above the waterline level, Porter built the gun deck. The beams on which the decking rested were bolted to the shield and supported by large iron knees. The knees had been left behind by the federals in one of the old shiphouses, and had survived the fire.

There was 7 feet of headroom between the gun deck and the deck that formed the top of the shield, called the "shield deck." The shield deck was around 16 feet wide, supported by 10-by-12-inch pine beams and covered fore and aft by a grating made up of 2-inch-square iron bars with a 2-inch-square mesh. The grating was designed to provide air and light to the gloomy gun deck, but the gun deck was still badly wanting for both.

At the very forward end of the shield sat the pilothouse, a low, cone-shaped structure made of cast iron about a foot thick. Brooke did not think much of it. "It will I think break into fragments if struck by a shot," he wrote.

There is some indication that there were actually two pilot houses, one on either end of the shield. A number of eyewitnesses describe two pilothouses, such as J. Tyler Jobson, who, in his quite accurate description of *Virginia*, refers to the "pilothouses, one at each end . . ."

Even Porter's son wrote, "Mr. Porter subsequently had two cast iron conical shaped pilot houses made and put on at each end." A contemporary model of *Virginia* currently in the Museum of the Confederacy in Richmond clearly shows two conical pilothouses.

A number of other observers and sketches made by eyewitnesses, however, show the vessel with only one pilothouse. These include a sketch made by Porter in a letter to a friend while construction was under way, making it more likely that she had just one. A photograph might answer this and a number of other questions about the ship's construction, but there is no known photograph of the *Virginia*.

A third of the way aft on the shield was the large smokestack for the four Martin-type boilers. Unable to telescope as some stacks were (John Ericsson had developed the first telescoping stack), it was enormously vulnerable to enemy fire. As it became riddled with holes during battle, the difficulty of keeping up steam in the boilers, described as "sluggish in draught," increased.

Virginia carried two small boats, probably hung on davits and resting on folding chocks mounted to the casemate. Neither these, nor the stanchions and railings surrounding the shield nor the anchors, survived long when the ship steamed into the first hailstorm of Yankee shot.

In the early days of *Virginia*'s construction, the navy yard was quite accessible to visitors, and the ship became a major sightseeing draw. According to Ramsay, "crowds of people, idle soldiers and others, used to assemble around the dry dock, and express their opinions concerning her, the predominant ones were, first, that she would go to the bottom as soon as she sank below the knuckle . . ." If she did not sink, it was assumed that the first enemy broadside would be the end of her.

One of the *Virginia*'s critics was an old friend of Ramsay's, Captain Charles McIntosh, who was in Norfolk waiting for orders. His last words to Ramsay, on the evening he left for a command at New Orleans, was, "Good-bye, Ramsay, I shall never see you again; she will be your coffin."

A little more than a month after the battle between *Virginia* and *Monitor*, which Ramsay survived, McIntosh was killed while in command of the ironclad *Louisiana* in New Orleans.

Along with being one of the great watershed events in American history, the Civil War remains the country's richest source of irony.

Chapter 17

"An Iron-Clad, Shot-Proof Steam Battery..."

The first battle involving an American ironclad took place on October 12, 1861. It was a middling success.

The ironclad had nothing to do with Ericsson, Mallory, or any of them. Initially, it had nothing to do with the navy at all.

The vessel was the *Manassas*, one of the great oddities of the war. She was originally a Massachusetts-built tugboat called *Enoch Train*. After the Confederacy began issuing letters of marque, the government licenses for citizens to sail as privateers, she was purchased by a New Orleans merchant, John A. Stevenson, for conversion into an ironclad. Her superstructure was removed and a rounded iron shell built on her hull and a massive iron ram mounted on the bow. *Manassas* looked like nothing so much as a partially submerged whale with a smokestack on its back and a single gun peaking out of a hatch near the bow.

The Confederate navy, seeing how useful the *Manassas* could be, commandeered her for the service. On the night of October 12, *Manassas* and a small fleet of *ad hoc* Confederate gunboats attacked the Union squadron at anchor at the Head of the Passes, a point on the Mississippi River about 90 miles below New Orleans.

Manassas, with a full head of steam and running with the current, rammed the flagship, USS *Richmond*. Unfortunately (or fortunately, depending on one's perspective) instead of making a direct hit, *Manassas* first hit a coaling schooner tied alongside, and did as much damage to herself as she did to the *Richmond*.

Still, the presence of the "infernal ram," whose existence had long been rumored, sent the Union sailors into a panic. Firing wildly into the night, they slipped their anchor chains and drifted as fast as they could for the open water of the Gulf.

The New Orleans *Daily True Delta*, in that wonderful hyperbole of nineteenth-century journalism, called the event "a complete success, and perhaps the most brilliant and remarkable naval exploit on record." It was hardly either of those things. Slow and unreliable, the ironclad frightened the Union sailors all out of proportion to her potential for mischief. In that regard, *Manassas* offered a glimpse of the extraordinary psychological effect that these new weapons, these ironclad rams, would have on people on both sides of the fight, military and civilian alike.

"Ram Fever"

At Newport News and Fortress Monroe, the Union's last toehold in Virginia, they had already begun to feel the effects of "ram fever." Perched at the tip of the peninsula between the James and York rivers, Fortress Monroe guarded the entrance to Hampton Roads from the Chesapeake Bay. As long as the Union held control of the sea and could resupply the massive fort and the troops at nearby Newport News by water, it seemed there was no Confederate force that could dislodge them. But now the Yankees were beginning to wonder.

As early as October 8, Gustavus Fox wrote to his wife from Washington, D.C., saying, "I went down to Old Point Sat. to see our new flag officer [Louis Goldsborough] . . . There is quite a panic down there from the ironclad vessels the enemy have . . ."

Two days before Fox's visit, the commander at Fortress Monroe, General John E. Wool, had written to Winfield Scott, saying that he believed the ironclad *Merrimack* would strike soon and that he could not be responsible for what might happen if he did not receive reinforcements. What Wool thought soldiers could do against an ironclad that he himself believed impregnable is unclear.

As formidable as CSS *Virginia* would prove to be, the panic being felt at Old Point and Fortress Monroe was almost half a year premature. As Wool was calling for more troops, Confederate Chief Engineer Williamson was only just completing the refurbishment of the ironclad's engines. But such was the fear of these newfangled "infernal machines" that even professional soldiers lost all sense of perspective, and they would again and again.

From the little intelligence he was able to receive from Norfolk, Gideon Welles might have had a better sense of how long it would be before the converted *Merrimack* was ready to attack. Though he lagged behind Mallory in his appreciation for ironclads, once he settled on a course of action he was eager to go.

As soon as the ironclad board had reached its decision regarding the shotproof, turreted battery, Welles told Ericsson to proceed with building the ship even before the ink on the contract was dry, indeed even before there was a contract at all. Incredibly, this was just eight days after Ericsson had shown the pasteboard model to Cornelius Bushnell.

A more astute businessman might have insisted on a written contract before he began placing orders for iron plate, but Ericsson was far more energetic than astute in such matters. The engineer returned to New York and without pause initiated a frenzy of ship design and building that would culminate in the most advanced naval vessel ever built up to that time.

Ericsson had assured the ironclad board that his shotproof battery could be built in a mere one hundred days. This was a major selling point. The Confederate ironclad building in Gosport was already hanging like the Sword of Damocles over the heads of the Navy Department, and they were anxious for something to counter it. *New Ironsides* was projected to take nine months to build. *Galena* would be done quicker, in four months, but that was still longer than the ironclad board wished to wait. Nor would *Galena* have been much help in keeping *Virginia* at bay. In battle, she would have been little more than an inconvenience for the Confederate ironclad.

The Resources of the North

Though Ericsson was confident that the ship could be built in one hundred days, he and his newfound partners knew it would not be an easy thing. In their favor were the enormous resources of the North, such as the Confederacy could not hope to match. In 1860, for example, the Southern states produced a total of 26,252 tons of wrought iron and steel to the North's 451,369 tons. But to finish the battery in one hundred days, the partners knew they would have to spread the work around, contracting different parts of the ship to different manufacturers.

Each of the partners agreed to handle a certain aspect of the project. John Griswold took charge of the finances. The Navy Department had agreed to pay $275,000 for the floating battery, doled out in $50,000 installments. The first $50,000 was due when the navy's superintendent of construction deemed that $50,000 worth of construction had been completed, meaning that the partners had to come up with the initial capital.

Griswold and Winslow secured the financing to get the ship started. (Years later, when he was running for governor of New York, Griswold would have his friend General Benjamin Butler make the claim on the floor of the Congress that he, Griswold, had financed the entire project out of his own pocket, only asking for repayment from the government after *Monitor* had fought the *Virginia*. Much of the valuable history of the *Monitor* project written by Gideon Welles was penned in order to refute that preposterous claim.)

Winslow arranged for the acquisition of iron. Though Griswold and Winslow were co-owners of the Rensselaer Ironworks, it was Winslow who was most interested in iron manufacture, while Griswold leaned more toward politics. Not surprisingly, Winslow turned first to his own company, as well as the larger Albany Ironworks, which he also managed.

Rensselaer and Albany Ironworks produced most of the ironwork for *Monitor*, save for the iron plate. For the plate, Winslow contracted with H. Abbott & Sons in Baltimore, Maryland, the only company outside of New York state to do significant work on *Monitor*.

Abbott & Sons had been founded in 1836. By the outbreak of the war, their 10-foot rollers for making iron plate were the largest in the country. One of the consequences of Maryland's not leaving the Union was the loss to the Confederacy of that resource, which, along with Tredegar, was one of the two largest ironworks in the South. As it was, while Abbott & Sons turned out iron plate for the Union *Monitor*, the Tredegar Iron Works, 130 miles away, was cranking out armor for the Confederate *Virginia*.

While parts and iron plate were manufactured in various places, all of the major construction of the Ericsson Battery* took place in New York City. Today one doesn't think of New York City as an industrial center, but in the mid-nineteenth century it was just that. In fact, New York City was one of the world leaders in the manufacture of steam engines; its closest rival was Glasgow, Scotland. It was a great convenience for John Ericsson that the *Monitor*'s three major builders were all within a few miles of his home.

There was no question as to which company would build the *Monitor*'s main engines and primary machinery: Delamater Ironworks. The owner of the ironworks, Cornelius H. Delamater, was Ericsson's closest friend and consistent business partner. It was Delamater who had inadvertently set the entire *Monitor* project in motion when he suggested Bushnell consult with Ericsson concerning *Galena*'s stability.

Ericsson's relationship with the ironworks, which was originally called the Phoenix Foundry, began soon after his arrival in New York. It was the Phoenix Foundry that built two of the iron canal boats for which Ericsson was commissioned by Robert Stockton and never paid. The Phoenix Foundry made the boilers, propeller, and blower for the *Princeton* and bored out the ill-fated *Peacemaker* under Ericsson's direction.

When John Ericsson first came to the Phoenix Foundry, it was a small collection of buildings at the foot of 260 West Street on the Hudson River. The foundry was owned by James Cunningham, a former builder of sailing ships. Cornelius Delamater was employed by Cunningham as a clerk. It was around this time that the two men met, and

---

*Until she was named, *Monitor* was generally referred to as the "Ericsson Battery" or the "shot-proof floating battery" or some similar combination. The use of the word *battery* is interesting, as it illustrates the fact that the builders did not quite know what they were building. Was it a ship, or some kind of floating fortress? The same was true of *Virginia*.

though Delamater was Ericsson's junior by eighteen years, a friendship was formed.

Three years later, Delamater and one of the foundry's engineers, Peter Hogg, assumed control of the business from the aging Cunningham. They expanded it, moved it to a new location on the East River, and renamed it Hogg & Delamater Ironworks. In 1858, Hogg sold his share of the foundry to his partner, and it became the Cornelius H. Delamater Ironworks, which it would remain until its demise long after the war.

In the two decades between Ericsson's emigration and the start of the Civil War, the little foundry was continually expanded, and its great success made Delamater a wealthy man. The Hogg and Delamater Ironworks became a sort of personal foundry for Ericsson. For his own projects, Ericsson was never charged for the use of any of the foundry's tools, facilities, or materials. In exchange, Ericsson brought all of his considerable business to Delamater. Delamater brought to the partnership the clear business sense that Ericsson lacked. Financially, the men were a boon to one another.

Personally, their friendship never faltered, though it was often strained by the pressures of business and Ericsson's quick temper. Delamater called Ericsson "John" and Ericsson called Delamater "Harry," intimacies almost unknown in Ericsson's other relationships. In October 1861, "Harry" Delamater agreed to build the main engines, boilers, propeller, and sundry other machinery for the Ericsson Battery.

Manufacture of some of the smaller mechanical systems was done by Clute Brothers Foundry in Schenectady, New York. For several years, Clute Brothers had produced Ericsson's small caloric engines. Now they were contracted to provide the gun carriages, the anchor windlass, the engine room grates, and the small steam engine that turned the turret.

The turret itself was built by the Novelty Ironworks, also situated on New York's East River. Novelty Ironworks was about thirty years old by the outbreak of the war. It was the largest manufacturer of steam engines in New York City, and one of the largest in the country. It was also, apparently, the only facility in New York City with steam presses and other equipment powerful enough to bend the turret's iron plates.

The hull of the *Monitor* was contracted to the Continental Ironworks, a shipyard in Greenpoint, in the borough of Brooklyn, across the

East River from Manhattan and the other two major builders. Continental Ironworks was only a year old in 1861, though its owner, Thomas Rowland, himself only thirty-one years old, had already built a number of vessels. At the same time that *Monitor* was being built, Rowland had two ships of more than 300 feet in length under construction.

Along with organizing the construction of the ship, Ericsson had another job that was at least as important: designing the vessel. The pasteboard model and the specifications he had shown to the ironclad board were leftovers from his pitch to Napoleon III. While the basic concept of a mostly submerged hull with a single turret remained the same, there were significant changes needed, and the working drawings from which the ship would be built had to be created.

Ericsson had to modify his original concept if the ship was to be built in one hundred days and for $275,000. The rounded hull and cupola-topped turret of the Napoleon model would never do. Shaping iron plate into those multidimensional curves was too costly and time-consuming. The new battery needed simple curves, and few of those. "[E]very part of the *Monitor* is straight or curves in one direction only," her engineer wrote, "no compound curves."

Ericsson set about his various tasks with his usual energy, described with Ericsson humility as "such herculean labor . . . as is not on record in the history of engineering." Meanwhile, the Navy Department was drawing up the actual contract for the ship.

In early October, with work on the battery already in full swing, Cornelius Bushnell returned to Washington to pick up the formal contract for signature. He was in for a nasty surprise. There was still a large contingent of naval officers who felt the Ericsson Battery would never work; "another Ericsson failure" was still the prediction. This faction was vocal with its concerns, and the ironclad board could not help but hear them, and in turn reevaluate their decision. The result was a cover-your-ass contract that placed an absurdly heavy onus on the contractors.

Much of what the contract contained was just what Bushnell expected. Ericsson, Winslow, Griswold, and Bushnell were contracted to build "an Iron-Clad, Shot-Proof Steam Battery of iron and wood combined on Ericsson's plan." It was to be 179 feet long and 41 feet on the beam. The ship was to be built of the best materials and employing the finest workmanship, though that was hardly something one need tell John Ericsson.

Bushnell may have been amused to find that the contract called for the builders to "furnish *masts*, *spars*, *sails*, and *rigging* of sufficient dimension to drive the vessel at the rate of *six knots* per hour in a fair breeze of wind." The ironclad board's original advertisement for designs had specified that each ship have masts and wire rope standing rigging, which both *Galena* and *New Ironsides* did have. Apparently the board felt *Monitor* should as well, though the idea of the *Monitor* with a sailing rig was absurd. In any event, Ericsson ignored that stipulation and nothing was ever said about it.

The United States agreed to pay the contractors the $275,000 that Ericsson had estimated it would cost to build the ship, less 25 percent held back in reserve. But the government had given itself more protection than a mere 25 percent. The contract specified that within ninety days of delivery, the navy would make a "test of the qualities and properties of the vessel." If the ship were to fail in any way, the contractors or their heirs would be bound to refund the entire cost of construction to the government. Such an arrangement would be risky for the contractors of any vessel; for one so utterly novel as the *Monitor*, it represented an extraordinary risk.

Bushnell already knew what the government meant by a "test." Commodore Smith had in mind something very specific, and it went well beyond the usual sea trials. For *Monitor* to prove herself, she would have to be tested under the fire of the enemy's guns. The navy would be willing to consider her a success if she survived.

Chapter 18

Hard Terms

The specific test of the *Monitor* that Gideon Welles and Commodore Smith had in mind was not spelled out in the contract. It was top secret. But the plan, in general terms, was already known to the contractors. On September 30, Smith had written to Ericsson, saying:

> So soon as the vessel is ready for service the Government will send her on the coast and put her before the enemy's batteries in the service for which you intend her. No other test can be made to prove the vessel and her appointments than that to which both parties agree to expose her . . . The plan is novel and because it is so, the Government requires the designer to warrant its success. Placing the vessel before an enemy's batteries will test its capacity to resist shot and shell—that is the least of the difficulties I apprehend in the success of the vessel, but it is one of the properties of the vessel which you set forth as of great merit. The Government cannot consent to receive the vessel until she shall have been tested in the manner proposed.

By "least of the difficulties I apprehend," Smith most likely meant that the vessel would, in his opinion, roll over and sink long before it got within range of the enemy's batteries. Gideon Welles, writing some years after the war, revealed the specifics of the navy's intended test:

> When the contract for Monitor was made, in October . . . the Navy Department intended that the battery should, immediately after reaching Hampton Roads, proceed up Elizabeth river to the Navy Yard at Norfolk, place herself opposite the dry-dock, and with her great guns destroy both the dock and the Merrimac. This was our secret.

This "test" was bad enough, but it was not until Bushnell read the contract that he discovered failure meant reimbursing the government their quarter of a million dollars. If *Monitor* could not stand up to the Confederate guns, the four contractors would be left with a useless, battered ship, perhaps even sunk or captured by the enemy, and a huge debt to the U.S. government. As Bushnell put it, "this seemed to me hard terms, but the life of a nation was at stake."

Bushnell took the contract and returned to New York, where he presented it to his partners. Ericsson, as was his nature, was unconcerned, in fact considered it "perfectly reasonable and proper." The contractors would only have to pay the money back if the battery proved to be a failure, and since that was not possible, there was no reason for concern. "If the structure cannot stand this test, then it is indeed worthless," he agreed. In fact, using the battery to destroy the captured ships at Norfolk was exactly the plan Ericsson had suggested to Lincoln when he wrote to the president in August.

Bushnell also claimed to have been satisfied with the contract's requirements, and was willing to sign.

Griswold and Winslow, who lacked the emotional involvement of the other two men, were not nearly so sanguine about this new development. Griswold was reluctant, but he agreed, provisionally. He told Bushnell that his signing the contract was contingent on Bushnell's finding "sureties," that is, the backing of people with sufficient capitol to pay the government in the case of disaster. This was also one of the government's requirements.

For Winslow, the contract was a deal breaker. He would not sign on to such an agreement.

Bushnell and the others gave Winslow a week to change his mind. In the meantime, Bushnell lined up a new fourth partner who was willing to step in if Winslow left. While Winslow dithered, Bushnell also arranged for two men of great personal fortune, N. D. Sperry of New Haven and entrepreneur Daniel Drew of New York, to stand as sureties for the project.

The question of the sureties is somewhat confusing. The contract that Bushnell and company signed specifically names *them*selves as sureties, and the United States District Attorney for the Southern District certified that the four men were "sufficient to pay any sum that may be demanded of them." But years later, Bushnell would state categorically that the four of them were *not* sureties, but simply the contractors.

Despite what it said on the contract, it was Sperry and Drew whom Bushnell later identified as the sureties for the *Monitor*.

Perhaps Bushnell, Ericsson, Griswold, and Winslow were the original sureties, but under the more onerous terms Griswold and Winslow would only agree to stay on as contractors if others could be found to take the position of sureties. Whatever the arrangement, Winslow finally came around, and on October 4 the contract was signed.

The Keel Is Laid

Three weeks later, on the 25th, a contract was finalized between the *Monitor* four and Thomas Rowland of the Continental Ironworks for the construction of the battery's hull. The contract made it clear that John Ericsson called the shots. It stipulated that Continental Ironworks "agrees to do the said work in a thorough and workmanlike manner and to the entire satisfaction of Captain Ericsson." If Ericsson did not feel the work was going fast enough, he had the authority under the contract to insist that Rowland hire more workers or have the men work longer hours. In exchange, Continental Ironworks would be paid seven and one half cents per pound of iron used in the hull's construction.

Monitor's keel was laid on the same day that the contract was, signed. Though the date of the keel laying is considered the official start of construction, quite a bit had already been done by that point. A new shiphouse, built specifically for the *Monitor* and paid for by the *Monitor* contractors, was already under construction at the yard in Greenpoint. Three weeks prior to the keel laying, Ericsson reported to Admiral Smith that "Mr. Winslow is now rolling the iron for the vessel itself, of a better quality, best snap iron, than has yet been put into any vessel in this country. The steam machinery for the battery is quite far advanced."

Also far advanced of the keel laying, and even advanced of the contract being signed, was the chronic buyer's remorse that Commodore Smith began to suffer.

The ironclad board had freely declared themselves ignorant of ironclad technology. They had been persuaded to build the Ericsson battery based on its creator's enthusiastic presentation. But now, having committed himself, and without Ericsson there to bolster him, Smith was not so sure.

By late September the old commodore began to pester Ericsson with a series of letters that grew increasingly pessimistic. On the 25th he wrote, "I am in great trouble from what I have recently learned, that the concussion in the turret will be so great that men cannot remain in it and work the guns after a few fires with shot."

Where Smith "learned" this is unclear, but it was no doubt from the cadre of Ericsson detractors who now had his undivided attention. Not the least of that group was Engineer in Chief Benjamin Franklin Isherwood, who considered Ericsson to be an unreliable eccentric. Isherwood's opinion would have carried a lot of weight with Smith, and must have greatly augmented his anxiety.

Ericsson, to his credit, was patient in replying to Smith's doom-and-gloom correspondence. Regarding the question of reverberation, he politely replied, "The difficulty you apprehend in relation to vibration of the air within the turret of my intended battery, I have most effectively guarded against." He went on to remind the commodore of his tenure as an artillery officer in the Swedish Army, which gave him ample experience with firing guns in close quarters. With the sides of the turret so thick that they could not vibrate and a grating above and below, the gunners would find their working conditions "a luxury" compared to the gun deck of a frigate.

Smith was mollified on this point, but there would be others. The old man was in a difficult position. As head of the ironclad board he was primarily responsible for selecting the Ericsson Battery. As chief of the Bureau of Yards and Docks, new ship construction was his responsibility. Gideon Welles pointed out that had *Monitor* been a failure, Commodore Smith, "more than anyone but the Secretary, would have been blamed, and was fully aware that he would have to share with me the odium and the responsibility." Worse, a lifetime of experience with wooden sailing men-of-war left him with no way to judge for himself the iron battery's chances of success or failure.

Ericsson in Action

Ericsson's efforts at this stage of the ship's construction were indeed herculean, even if he did say so himself. He spent his days visiting the three construction sites: Delamater Ironworks, Novelty Ironworks, and Continental Ironworks, the third receiving the bulk of his attention. Ericsson was fifty-eight years old the year *Monitor* was built. Thomas Rowland, almost half Ericsson's age, was astounded by the engineer's energy. "Mr. Ericsson was in every part of the vessel, apparently at the same moment, skipping over planks and gangways, up and down ladders, as though he were a boy of sixteen. It seemed as though a plate could not be placed or a bolt struck without his making his appearance at the workmen's side."

For Rowland, too, it was a herculean task of organization and management. Hundreds of men were hired to work on the battery. Shifts were often worked around the clock. Not only was the pace of work faster than usual, but the vessel was utterly unique, so that virtually every part of her construction was a novelty, with plans, parts, and materials flowing constantly into the shipyard.

At the end of the day, when Ericsson was done at Continental, he returned to his office at 95 Franklin Street and designed. Late into the night he sat at his desk and drew up the plans for the shotproof iron battery, even as the ship itself was being built at scattered venues around the state. It was an extraordinary juggling act. Ericsson had to make certain that Winslow knew what type of iron plate or bar or bolt was needed, where and when it was needed, and had to make certain that the builders at the other end knew what to do with the iron as it arrived, even as the vision of the ship was just flowing from his mind onto the working plans.

That Ericsson could produce drawing after drawing, totaling more than one hundred in the end; send them off to the disparate manufacturers for fabrication; and do so with such precision that when, at last, all the parts were assembled, not one of them needed modification, is testament enough to his genius. This would have been an extraordinary feat even for a ship of proven design, and the *Monitor* was anything but that. By the estimate of her engineer, Isaac Newton, *Monitor* contained at least forty patentable inventions, though Ericsson, so often casual about his intellectual property, never bothered to patent them.

As work progressed, Commodore Smith grew more worried. Two weeks after his letter regarding the firing of guns in the turret, he wrote, "Computations have been made by expert naval architects of the displacement of your vessel, and the result arrived at is that she will not float with the load you propose to put upon her, and if she would, she would not stand upright for want of stability, nor attain a speed of four knots."

Even as Ericsson was designing the ship, others were evaluating what little they knew about his work and finding it wanting. In this case the computations were most likely made by the navy's chief constructor, John Lenthall, a capable but old-school naval architect who referred to the floating battery as "Ericsson's iron pot." The men building the *Monitor* and those building the *Virginia* would have had a great deal to commiserate about regarding the absolute lack of faith in their respective ships.

In the same letter, Smith suggested a fix—filling the angle between the lower and upper hull with wood—that would have made Ericsson grit his teeth. It must have been vexatious in the extreme to have someone with no knowledge of naval architecture make suggestions for unnecessary fixes.

Three days later Smith wrote to Ericsson to inform him that he, Smith, had calculated the battery's displacement and concluded that it would not float. He also touched on the real reason for his anxiety. "I shall be subjected to extreme mortification," he wrote, "if the vessel does not come up to the contract in all respects . . . I assumed a great responsibility in recommending in haste (to meet the demands of the service) your plan."

As if he did not have enough to do, Ericsson wrote back a week later, giving the Commodore a long dissertation on stability. Whether or not Smith could follow the mathematics of Ericsson's proof, he was still not entirely convinced that the vessel would float. However, with ship-yards, foundries, ironworks, and John Ericsson all going full bore to get the vessel finished, there was nothing he could do at that point except wring his hands and write letters.

The "impregnable battery" was well under way, and the only real question left, the one that really plagued Gideon Welles and Gustavus Fox and Joseph Smith, was whether she would be done before the *Virginia*.

Chapter 19

Many Vexatious Delays

As eager as Welles was to complete *Monitor*, Mallory was just as eager to finish the *Virginia*. He considered it the highest priority.

On August 18, he wrote to Flag Officer Forrest, reiterating the need for haste. "The great importance of the service expected from the *Merrimac*, and the urgent necessity of her speedy completion, induce me to call upon you to push forward the work with the utmost dispatch." He told Forrest that Porter and Williamson were to receive "every possible facility at the expense and delay of every other work on hand . . ."

There seemed to be no shortage of men to work on the ship, but skilled men were another matter. Catesby Jones pointed out that most of the "vexatious delays" in the ship's construction "arose from the want of skilled labor and lack of proper tools and appliances."

H. Ashton Ramsay would have seconded that sentiment. "I was greatly surprised," he wrote, "to see how little the Confederate Authorities seemed to value the navy yard." He complained that the yard had been stripped of everything that the army might find useful in the way of tools and materials, that mechanics had been discharged, and that those who were kept had their wages cut.

Ramsay was already aware of the Union's great advantage in manufacturing, and his short and unhappy stay in Boston had impressed upon him their determination to wage war. If his fellow Confederates could see what he had seen, he felt, then "the yard should be full of mechanics" and work would progress without rest.

The culprit, in his opinion, was the prevailing belief that England and France would join the war on the Confederate side and the United States would quickly sue for peace. Nor was the navy's situation helped by the recent victory at Manassas. "The people were full of the prowess of the army," he wrote, "but were unreasonable concerning the navy."

Still, the work progressed, and as summer gave way to fall the casemate was very nearly completed. Flag-officer French Forrest, commandant of the yard, became increasingly obsessed with security, fearing, and not without justification, that there were many secret Union loyalists still employed there. By August, the throngs of sightseers who had earlier gathered in the shipyard to watch the progress were halted, and no one was allowed in without a written pass from an officer.

Later that month the pass system became official with Forrest's general order "to prevent thoughtless communications [loose lips sank ships even in the Civil War] or disloyal persons from visiting this Dockyard."

By the middle of October, Forrest took the next step, administering a loyalty oath to all dockyard workers. On the Holy Bible, workers swore they would be "truly loyal in support of the Southern Confederacy." But still information leaked North.

Porter and Brooke

If Brooke had furnished the concept of *Virginia*, Porter definitely filled in the details. Because of her unique construction, many of her inboard fittings, things that would have been no more than afterthoughts on a conventional ship, such as capstans, hawse pipes, and steering gear, had to be carefully engineered.

Adding to the construction delays were changes suggested by Brooke, such as the addition of four more gun ports. Porter had cut just one gun port each for the bow and stern guns, allowing them to point only straight ahead and straight astern. An enemy approaching either bow at an angle, or coming at the ship's quarter, would be out of the line of fire of any gun. Brooke suggested piercing those blind spots of the shield so that the bow and stern guns, on pivots, could be run out in any of three directions. This change was made. Porter's original ink drawing of the "*Merrimac* Gundeck," in the possession of the Mariners' Museum in Newport News, Virginia, clearly shows the additional gun ports sketched in in pencil.

Brooke also suggested increasing the number of hatches from two to four, which Porter did. Brooke objected to the manner in which the wheel ropes (actually chain) were run from the rudder in through the casemate, claiming "they were liable to be jammed by a shot." This Porter did not change until later, when Catesby Jones also objected to the arrangement.

Porter was stretched to his limits. Just as Mallory was relying more and more on Brooke to handle a variety of ordnance and design functions, so Porter was quickly becoming the *ad hoc* chief constructor of the navy. In early October, for instance, with *Virginia's* construction in full swing, Porter was instructed "to commence laying the keels of the two gun Boats ordered by the Sec'y of the Navy."

"I never was so busy in all my life," Porter wrote to a friend, "I have all the work in the navy yard to direct, and all the duty of the Bureau of Constructor of the Navy to attend to besides."

Porter was not then officially the navy's chief constructor; in fact, the office did not exist. The following year, when Porter found that Congress was planning to establish the office of chief engineer, he wrote to Mallory saying, "I would respectfully request that you ask Congress to establish the office of chief naval constructor of the Confederate States, the duty of which I have been performing for some time." He further suggested that he, Porter, be given the position. Mallory did as Porter asked, and Porter became Chief Constructor of the Navy. William Williamson was given the position of Chief Engineer of the Navy.

Williamson was in a situation similar to Porter's. Though his primary concern was *Virginia*, his attention, at Mallory's insistence, was constantly divided among other projects. The secretary was growing increasingly enamored of ironclads on the *Virginia* model. By the end of the year he would all but abandon his hope of having a *La Gloire* or a *Warrior*. Mallory was dissuaded by the great expense and the fact, as he

now understood, that such a vessel could not be built in a private shipyard, and that the governments of England and France would not help
at the risk of violating their neutrality. There was no possibility of building such a ship in the Confederacy.

Ironclads along the lines of *Virginia* were another story. "Such vessels as those . . ." he told Jefferson Davis, "can be constructed in a third
of the time which would be required to build a sloop like the *Gloire* if
we had the ability to build one." By October 1861, Mallory had four
substantial ironclads under construction, not including the *Virginia*.
More were planned.

By October, Chief Engineer William Williamson had completed the
work on the *Virginia*'s engines and his talents were needed elsewhere.
Mallory sent him to New Orleans to help the Tift brothers, Asa and
Nelson, who had witnessed the shield tests on Jamestown Island, with
work on their ironclad, the *Mississippi*. The old *Merrimack*'s much ma-
ligned engines were now the sole responsibility of H. Ashton Ramsay.

Iron Plates and Flats

By early November the shield was done as well, at least the wooden
part, pine and oak 24 inches thick with caulking between all the planks
to render it watertight. *Virginia* was an ironclad without any iron.

The first plates, 1 inch thick, had been rolled in September, before
Brooke had performed his tests and determined that 2-inch plate was
needed. The already finished 1-inch plate, less than a third of the total
armor, was sent to Gosport and used on the ship, "criss-crossed on the
sides of the shield to make up the four inches." The foredeck and the
afterdeck as well were armored in 1-inch plate.

The armor below the waterline was also made up of 1-inch plate,
though Catesby Jones claimed that "it was intended that it should have
been three inches." This belt of iron, running around the circumference of the ship, extended down from the eaves of the shield two or
three feet. After the battle, this was changed to three and a half feet.

The rest of the plate, more than two-thirds of it, was 2 inches thick,
drilled for bolt holes and redrilled and redrilled again as Brooke and

Porter changed the configuration. By October, Tredegar was going full bore, "pressed beyond endurance," as one of the ironworks' partners wrote, producing plate for the *Virginia*.

As Tredegar produced plate, the problem became getting it to Gosport. "The Navy Department is pressing us to send forward the heavy iron for the steamer *Merrimac*. We have some 70 to 100 tons of iron now ready to ship," one of Tredegar's people wrote to the superintendent of the Richmond and Petersburg Railroad. "Will you please state if you have a sufficient No. of flats to forward the iron promptly."

But the railroad did not have enough "flats," or flatcars. Southern railroads were not much to brag about before the war, and they were getting increasingly worse. An endemic shortage of rolling stock was now exacerbated by military demands. Tredegar was shipping its heavy guns all over the South, and the flat cars were not coming back as quickly as they were going out. What flats there were were often needed to transport supplies to the army, still facing off with the federals across the Potomac River 80 miles away.

Mallory may have considered the *Virginia* a top military priority, but not everyone did. At one point, iron on its way to Gosport was unceremoniously dumped off a flatcar to make room for other war matériel. Flag Officer Forrest had to dispatch a man to search the tracks from Gosport to Richmond, find the iron, and load it on another flat.

By mid-November, one of the Tredegar partners would write, "We have iron for the Navy Yard that has been lying on the bank for 4 weeks—several sizes are ready to go down, the Rail Road has been unable to transport it." To alleviate the transportation problem, French Forrest arranged to move iron plate along various different routes and rail lines. He instructed his agent, William Webb, to arrange for shipping *Virginia*'s iron by way of Weldon, North Carolina. Using another railroad speeded delivery somewhat but doubled the distance the iron had to travel.

Most of *Virginia*'s armor was produced in November and December 1861 and January 1862. As it flowed into Gosport from various routes, it was mounted on the huge wooden frame of the shield. The first course, or layer, of plate was put on lengthwise, parallel to the ship's waterline. The next course, the outer layer, was put on vertically.

Iron bolts, $1\frac{3}{8}$ inches wide and around 30 inches long, were pounded through holes drilled in the shield that corresponded to the holes in the iron plate. The bolts were countersunk in the iron plate to leave a smooth surface on the shield. On the inside they were fastened with heavy iron nuts and washers.

In the end, Tredegar rolled and shipped 723 tons of plate iron for *Virginia*. For that they were paid $123,015, mostly in Confederate bonds.

The iron plate was the final part of the shield, but completion of the shield did not mean the *Virginia* was complete. She still needed port shutters, guns, gunpowder, oil, coal, officers, and crew, none of which were easily found in the South, with the exception of officers. Nor was the iron plate going on as fast as Mallory and the others would have liked. By Christmas 1861 only a small part of the casemate's starboard side was plated.

What's more, the Union had begun their own ironclad, and the Confederates were well aware of it. If they had any questions, they had only to read the *Scientific American* of November 28 (which no doubt they did). In that issue was a description of *Monitor*, largely provided by Ericsson, which was better than anything they could have hoped to get from a Confederate intelligence agent.

Anxiety was high and getting higher, both North and South. It was the Confederates' head start versus the Union's superiority in industrial capacity, and the race was well under way.

Chapter 20

The Ericsson
Battery

The "Iron-clad, Shot-proof Steam Battery of iron and wood combined" was an extraordinary creation. Its design represented the culmination of the many years Ericsson had spent grappling with the issue of an impregnable battery, combined with the need to build it fast and relatively cheaply. It was not flawless, far from it, and its flaws nearly destroyed it on its hellish maiden voyage down the coast. But given the fact that it was a prototype, and given the whirlwind of design and construction that brought it to life, *Monitor* was closer to perfection than anyone could have reasonably expected.

There were several considerations that most influenced Ericsson's design. The first was the size of the converted *Merrimack* and the extent to which she was already completed by the time construction of his own battery could begin. Ericsson knew that there was no chance of building a vessel of similar size in time to meet the Confederate ship. What's more, *Merrimack* would clearly be a powerful enemy, too powerful for any conventional vessel, including ironclads of the *Galena/ New Ironsides/Warrior* variety. What was needed was something radical and thoroughly shotproof.

Ericsson also considered the waters in which his battery would have to operate. Southern waters were shallow, and the channels in which vessels could navigate narrow. What an enormous advantage it would be if the guns could be pointed without having to point the entire ship!

The Confederates, of course, were perfectly aware of the navigational difficulties—it's why the deep-draft hull of the *Merrimack* was not Porter and Brooke's first choice. But their limited industrial capacity left them with few options, whereas Ericsson had open to him all the options offered by the industrialized North.

Ericsson also considered that Southern artillerists onshore would be able to lay their guns with such accuracy that any vessel not entirely shotproof would be torn apart. Those same shore batteries would make it very hazardous for any crew member working on an exposed deck. However, since Ericsson stubbornly refused to mount the masts and sails for which Welles called, there was virtually no work that needed doing on deck, save for working the anchor. That problem Ericsson neatly solved.

The Upper Vessel

Ericsson viewed the shotproof iron-clad battery as essentially two ships, one on top of another, so much so that they were generally referred to as the "upper vessel" and the "lower vessel." The upper vessel was 172 feet long, 41 feet on the beam, with a depth of 5 feet. It was, in essence, a huge wooden raft, covered over in iron plate, with a hollow underside that was designed to fit like a lid over the lower vessel.

The deck of the upper vessel was originally intended to be 6-inch-thick oak, but a lack of seasoned oak led Ericsson to substitute 7-inch pine planks on top of 10-inch-square oak beams with 26 inches between them. The top of this deck was covered with two layers of ½-inch-thick plate iron.

There was no bulwark or rail of any sort around the ship's perimeter. From the edge of the deck, the sides extended down 5 feet. The bottom 3½ feet would be submerged when *Monitor* was floating on her waterline, with 18 inches of freeboard showing above water. All of the ship's wetted surfaces were painted with red lead over a white zinc base to ward off marine growth.

The sides were built up of white oak, 26 inches thick. On top of the oak, forming an armor belt around the vessel, was 5 inches of plate iron built up out of 1-inch laminates. The beams and sides were supported by angle iron brackets and iron braces.

Near the bow of the upper vessel was a unique Ericsson design, a completely enclosed anchor well. It consisted of a 5-foot-diameter hole through the upper vessel, covered with a removable 2-inch-thick plate-iron cover. The anchor, a grappling-hook type designed by Ericsson specifically for the *Monitor*, was raised by means of a manually-operated windlass inside the ship. The anchor chain ran out a hawse hole in the after end of the anchor well, over a roller, down to where it was attached to the anchor. When raising the anchor, the men cranking the windlass stood safely in the confines of the ship, right below the pilothouse, while the anchor was pulled up into the protected well. From the outside, one could not even tell the anchor was being raised.

Ten feet back from the anchor well stood the pilothouse. Ericsson had originally envisioned a very spacious affair, a round house 5 feet high and 6 feet in diameter. That certainly would have been more accommodating than the pilothouse that was eventually placed on the bow, which was a minuscule 3 feet 6 inches long by 2 feet 8 inches wide on the inside and rose only 3 feet 10 inches above the level of the deck. It was well that he altered his plan. The tiny pilot house that ended up on *Monitor*'s bow caused enough problems in aiming the guns—the huge house first envisioned would have been a nightmare.

The pilothouse was cramped, to say the least. After the battle, a friend of Worden's who had toured *Monitor* wrote to ask if Worden alone had been steering the ship during the fight. "I am sure from my cursory observations," he wrote, "that *two* could not stand in that 'hole.'" In fact, a minimum of three did: the officer in charge, a pilot, and the quartermaster at the helm. A grated deck suspended 3½ feet below the level of the deck gave 7 feet of headroom to the men squeezed into that little box.

The pilothouse itself was built of blocks of solid wrought iron, 12 inches deep and 9 inches thick. The whole thing was held down by 3-inch bolts that passed through the corners, then through the deck and the beams below. The roof of the pilothouse consisted of a 2-inch thick iron plate that sat in a groove on the top of the house but was not bolted in place, so that in an emergency it could be pushed aside to allow the crew to escape.

In order for the pilot, quartermaster, and watch officer to look out, spacers were placed between the first and second iron blocks, creating a long slit around the circumference of the house. The original slit was ⅝ of an inch wide, which, as Ericsson explained, "affords a vertical view 80 feet high at a distance of only 200 yards." The navy men overseeing the battery's construction apparently felt this was not sufficient, and increased the height of the slit, a decision that Ericsson blamed for a number of the problems the ship encountered, including the water that poured in with enough force to "knock the helmsman completely round the wheel." (Ericsson went to great lengths to explain how anyone who thought there were any mistakes made in *Monitor*'s design were, in fact, themselves mistaken.)

At the stern of the upper vessel was another well, this one to accommodate the tops of the propeller blades as they rotated past. Like the anchor well, the propeller well was covered with a 2-inch plate iron cover that could be removed, thus allowing the propeller to be maintained or even removed with the ship still in the water.

Beyond the pilothouse and the turret, the deck of the *Monitor* was largely featureless. Behind the turret were the opening for the smoke pipes, and behind them the air intakes, the arrangement that had caused so many problems. These were fitted with flat "bomb-proof gratings" for battle, leaving the deck completely flush. There were also six 16-inch bunker plates covering the cylindrical chutes down which coal was dumped into the bunkers, as well as round skylights that illuminated the lower deck with "light enough for all purposes except reading and writing." The skylights could be covered with iron deadlights in battle.

The Lower Vessel

The lower vessel was 50 feet shorter than the upper vessel and 7 feet more narrow. It was essentially a flat-bottomed steamer with no deck or upper works, the deck being formed by the upper vessel sitting on top of it. The lower vessel was lightly built, but it could be, because it was so well protected by the upper vessel. Flat-bottomed, with sides that sloped up at an angle of 36 degrees to the horizontal, it was built entirely of "the best American plate iron ½ inch thick all over."

Part of the genius of the lower vessel was the sloping sides that kept that part of the ship from harm's way. Because the upper vessel overhung the lower by nearly 3 feet at its narrowest place, it was virtually impossible for shot to reach that thin iron hull. The article in *Scientific American* points out that the upper vessel "brings the lower body within such angles that shot cannot strike without first passing through water for a distance of more than 25 feet, and then striking at a very acute angle, 100 [degrees] at the most; while the propeller, rudder and anchor cannot be reached by shot at all."

The very forward end of the lower vessel, which formed the interior part of the ship, met the upper vessel just behind the anchor well. Furthest forward was the windlass for raising the anchor, and then the ladder and floor of the pilothouse. Directly behind the pilothouse, and spanning the width of the vessel, were the captain's quarters, with his cabin on the port side and his stateroom to starboard. Each of these rooms was around 10 feet square.

The captain's cabins opened onto a short alleyway that led to the wardroom, the officers' communal area, which was approximately 10 by 15 feet. On either side of the wardroom were tiny, closetlike cabins made of black walnut for the officers. The cabins contained a berth with four drawers below and small closets above. In one corner was a small desk, and diagonally opposite a washbasin equipped with a whiteware slop jar, tumbler, water pitcher, and soap dish, all with the name *Monitor* in gilt letters. There was eight feet, six inches of headroom under the beams. For the sake of ventilation, the walls of the cabins did not extend all the way to the overhead, which meant that the cabin walls did little to keep out noise.

Small as they were, the cabins were nicely appointed. According to one officer, "Capt. Ericsson fitted our rooms up at his own expense & has been very liberal. I have been on board nearly all the vessels that have left the Yard since I have been here & have seen no room as handsomely fitted up as ours."

A wooden bulkhead separated the wardroom, "officer's country," from the crew's quarters, or berth deck, aft. In this case, "quarters" meant simply a big open room, 16 feet wide by 25 feet long, in which the off-watch crew could congregate or hang their hammocks for sleeping. It was a small space for the forty-nine men who had to live there, but Ericsson figured that half of the crew would be on watch at any given time, and thus not on the berth deck. Two ladders on the centerline of the vessel led to hatches, one opening onto the deck, one leading into the turret. Storerooms lined either side of the crew quarters, and the galley and galley stove were situated at the after end.

Initially there was concern that the men's health could not be maintained if they lived most of their time below the waterline, as they would onboard *Monitor*. An 1864 report by the navy's Bureau of Medicine and Surgery, however, found that there was less illness aboard the monitor-type* vessels than aboard wooden vessels of comparable size.

Ericsson credited the monitors' healthy environment to his efficient ventilation systems, and no doubt that played a big part. The steam-driven blowers that provided draft for the boiler furnaces, pulling in 7,000 cubic feet of air per minute, also pushed fresh air into the crew quarters. The fresh air came in near the bottom of the hull, through registers in the deck that could be opened or closed, forcing the old air up and out. The downside, of course, was that if the blowers failed, the boiler fires died as the oxygen in the air in the space below decks was converted to carbon dioxide by the fires in the boilers. Consequently the below deck compartments filled with a deadly mixture of carbon dioxide and nitrogen, as the men of the original *Monitor* discovered on their first ocean voyage.

Eight inches forward of the center of the turret, which was also the center of the ship, was a ⅜-inch plate iron bulkhead. This bulkhead separated the ship into two sections and helped support the weight of the turret. All spaces forward of the bulkhead were accommodations for the men or storage for supplies and ammunition. Behind the bulkhead were the two coal-burning fire tube box boilers, built by the Delamater Iron Works, the engine, and the coal bunkers.

The two cylinder, trunk piston, vibrating lever engine was, like most of the ship, a unique Ericsson design. Unlike most of the ship, however,

---

*Following the success of the original, the word "monitor" became a generic term for all of the subsequent turreted, iron-clad vessels designed by, or inspired by, John Ericsson.

it was not designed specifically for *Monitor*. Ericsson had already employed an engine of the same design in two other vessels, the *Judith* and *Daylight*. Considerable time and risk were saved by using a proven design, and as it happened that unique engine was ideal for the *Monitor*. The engine had two 36-inch diameter pistons that moved in what was essentially one long single cylinder, separated into two separate cylinders by a diaphragm in the center, cast in place. The cylinders were set at right angles to the centerline of the ship and were connected to large diameter hollow trunk piston rods which ran through the cylinder heads. A steam gland at each cylinder head prevented excessive steam leakage.

The piston rods were in turn attached to a complicated series of rods, called vibrating levers and jackstafts, which ultimately communicated the engine's power to the propeller shaft. Valves mounted on the top of the cylinder casting were used to regulate the steam flowing to the cylinders as well as the direction in which the engine was turning, forward or reverse.

The Ericsson vibrating lever engine was a complicated affair. However, the experienced machinists at Delameter Iron Works, working under the close supervision of Ericsson himself and using the best materials available, were able to make it reliable and efficient. The Ericsson engine may have looked like a spastic iron grasshopper while it was running, with its long vibrating levers moving back and forth, but it worked, and had a very low profile, perfect for so shallow a vessel. The navy liked it, and the same type of engine was used in many of the subsequent monitors, and came to be known as the "monitor type."

The Turret

Perhaps the most unique aspect of *Monitor* was the turret, though even Ericsson was quick to admit that it was not an original concept. Ericsson did not invent the turret, nor did he claim to, and felt that any well-informed naval artillerist should be familiar with turret design. The Confederates certainly understood the idea of a gun on a revolving platform. In May 1861, the Confederate navy built "a 'turn table' for the purpose of mounting a XI inch gun on Craney Island."

Ericsson himself had first learned of the turret concept from an instructor, in fortification and gunnery around 1820, when it was already a well-known idea.

To Ericsson's knowledge, the first account of a "movable turning impregnable battery" was introduced by a Scotsman, Gillespie, in 1805. In 1807 an American, Abraham Bloodgood, published a picture of a circular, revolving floating battery in the *Transactions of the Society of the Promotion of Useful Arts in the State of New York*.

In 1843, another American, Theodore R. Timby, patented a revolving turret that was land-based. Timby tried to sell the idea to the government, but those who reviewed the design considered it too expensive and not particularly effective. Timby's patent, however, did give him a valid claim to the turret concept. After *Monitor's* success, and the government's subsequent order for more monitor-type vessels, Ericsson's associates agreed to pay Timby a $5,000 royalty for every turreted vessel they built, the total payment not to exceed $100,000. Ericsson never agreed with this arrangement.

One of the foremost claimants to the turret's development was the British naval captain Cowper Coles, who had been advocating the idea for some time. Stephen Mallory, looking back to his time on the Senate's naval committee (and perhaps with a touch of sour grapes), said, "Coles is entitled to the paternity of the *Monitor*. I studied his views attentively in 1855, and again in 1859." Welles was also aware of Cole's turret, and made some allusion to it, prompting Ericsson to reply that he had "perfected this invention more than seven years before Captain Coles brought out his abortive scheme in England."

John Ericsson may not have originated the idea, but he was the first to put it into practice. The turret that he designed for *Monitor* was 9 feet high and had an inside diameter of 20 feet. It was built up out of eight layers of 1-inch thick plate iron, with an additional layer around the gun ports, bent to a gentle curve by the presses at the Novelty Ironworks. The roof of the turret was made up of 2-inch-thick wrought-iron beams set 6 inches below the upper edge and supported underneath by 6-inch wrought-iron beams. Space left between the beams making up the roof provided air and light to the turret's interior.

The roof was pierced by two hatches with sliding covers. A series of stanchions could be installed in sockets around the perimeter of the turret. Holes in the stanchions allowed for a rope railing to be rigged. The stanchions also supported a canvas awning, making the turret roof the only sheltered area abovedecks, and the only place one could go topside in any kind of a seaway without fear of being washed away. Even at anchor, in bad weather the deck was treacherous. Fireman George Geer wrote to his wife while anchored off Fortress Monroe, "It has rained and blowed steady all the time, and the waves wash over us so it [is] most impossible to stand on Deck." Stanchions and awning were removed when the vessel was cleared for action.

Near the base of the turret, Ericsson placed four massive wrought-iron crossbeams running parallel to one another and stretching from one side of the turret to the other. The beams filled several functions. Foremost, they acted as slides on which the Ericsson-designed gun carriages would run in and out. The carriages rested on brass rollers that eliminated the need for large crews of men straining on tackles to run the guns out. On an even keel, one man could run out the massive gun by himself.

Ericsson intended for *Monitor* to mount two 15-inch Dahlgren smoothbore guns, the recoil of which would be checked by friction mechanisms "similar to that applied to the United States Steam Ship *Princeton*." The slides would allow the guns to recoil a maximum of 6 feet, while the friction mechanisms could reduce the recoil to as little as two.

The turret was pierced with two gunports, side by side, both guns aiming in the same direction. The ports were oval-shaped and much smaller than the square gunports of ships of the wooden navy. Since the aiming was done with the turret, not the guns, there was no need to move the guns transversely, which could not be done in any event, and only minimal need to adjust the elevation, since *Monitor* was intended to fight at close range (when she was later called upon to fire at elevated shore batteries on the James River, her guns could not train high enough).

Massive port stoppers, shaped like elongated teardrops, hung like pendulums behind the gunports. Eight inches thick and made of wrought iron they were intended to ward off shot while the men reloaded the guns. To that end, the stoppers were pierced with holes through which the rammer and sponge handles could be thrust while the stoppers were closed. The port stoppers were swung out of the way with a block and tackle, but in practice they proved very awkward, requiring every man in the turret hauling on the tackle to move them. This was not a problem in Ericsson's concept, since the port stoppers should have rarely been used. Rather, the turret should have been rotated away from the enemy while the guns were loading, and rotated back to fire.

The floor of the turret was made up of an iron grating. Two grated hatches in the floor could be made to line up with any of four hatches cut through the main deck to allow access from the interior of the vessel into the turret. A track running around the perimeter of the turret held the massive, 165-pound round shot that was hoisted by means of block and tackle into the muzzles of the Dahlgren smoothbore guns.

The entire turret sat on top of a vertical, 9-inch-thick iron axle. The top of the axle fitted into a socket mounted on the underside of the gun carriage beams, and the bottom was supported by a Y-shaped truss. This truss was riveted to the berth deck side of the ship's central bulkhead and ran from the floor beam to the overhead.

In action, the turret was "keyed up," raised into position by a large wedge with a threaded rod on one end. By tightening a nut on the wedge, the shaft was raised so that all of the turret's weight rested on the axle.

Below the turret and the main deck, the axle was connected by a series of gears to a two-cylinder steam engine. The controls for this engine were run up into the turret itself, so that the entire turret, weighing well in excess of 100 tons, could be controlled by one man. It took approximately twenty-three seconds to spin the turret 360 degrees.

When not in the raised battle position, the turret rested on a bronze ring set into the deck. The weight of the turret on the smooth bronze formed a watertight seal. There was no need for caulking, or so the *Monitor*'s crew learned.

Monitor was a revolution, an unprecedented creation springing from the mind of a single genius. "Ericsson's Folly," "Ericsson's iron pot," the "cheese box on a raft," *Monitor* was destined to change forever the course of naval ship design. But as the fall of 1861 gave way to winter, there were not a dozen men among all those involved in her construction who even thought she would float.

Chapter 21

"Do You Really Think She Will Float?"

By November 1861, Stephen Mallory and Franklin Buchanan, head of the Office of Orders and Detail, began to think about manning the *Virginia*. There was no shortage of bold young officers in the naval corridor between Norfolk and Richmond, and they all wanted to be onboard the ironclad. Buchanan would later write, "The ship is crowded with officers and those who belong to her cannot be very comfortable."

As early as September, rumor was circulating that Franklin Buchanan was to be captain, a position that French Forrest and other senior officers were also angling for. Lieutenant Robert D. Minor wrote to his friend Catesby Jones, "Buchanan will probably be her captain, and I hope you will be her first lieutenant."

Minor was thinking of himself as well. In another letter to Jones he wrote, "No detail yet for the *Merrimack*. I hope to be one of her lieutenants, and I need not say how pleased I would be to serve with you. Brooke, too, will join her, I suppose. He is an indefatigable fellow, and works with heart and head for our glorious cause."

John Mercer Brooke wanted very much to be part of *Virginia*'s crew. It was a chance for the kind of glorious action that naval officers crave. What's more, he feared that his staying ashore would imply a lack of confidence in the vessel he himself had designed, a real blow to the already shaky faith that most observers had in the ship.

Unfortunately, Brooke was swamped with ordnance work, work that was crucial to the Confederate war effort, even if it was not as dramatic as sallying forth in an experimental ironclad. Mallory was reluctant to let Brooke go. It would not be easy to spare him, even for a short time. If the ship was to be an efficient fighting machine, officers and men would have to drill together well before they got under way. For Brooke, that meant a lot of time away from his ordnance work.

Finally, Mallory relented. If Brooke could make arrangements for the Ordnance Office to function without him, he could sail on *Virginia*. It was what Brooke wanted, but he was not sure he could pull it off. "I do not now see clearly how I am to manage," he wrote in his journal.

In the end it was a moot point. Around the time *Virginia* was to get under way, Brooke's family was taken sick. Mallory called Brooke to his office and suggested that under the circumstances Brooke should not go out on the ship.

Brooke demurred, but in the end told Mallory that he would not go with the ship if Mallory thought he should not. "He [Mallory] said he did think so most decidedly," Brooke wrote. John Mercer Brooke would achieve lasting fame as a scientist, ordnance expert, and designer, but not as a fighting sailor.

Robert Minor's hope that Catesby Jones would be named first officer was fulfilled in November. On the 6th, Buchanan wrote to Jones, "You are hereby detached from the command of the Jamestown Island batteries, and you will report forthwith to this office [Orders and Detail] for duty." Jones had his good friend Brooke to thank for the dream assignment. Brooke had recommended the appointment to Mallory.

Powder and Shot

Because of Jones's expertise with naval gunnery, he became both *Virginia*'s executive officer and her ordnance officer. With the first Brooke rifles intended for *Virginia* coming out of the ironworks at Tredegar, Jones was ordered to "ascertain by actual firing their range and capac-

Bird's-eye view of Hampton Roads, arguably the most important stretch of water for either side during the Civil War. Fortress Monroe and the Rip Raps commanded the entrance to the Roads, which led to the Elizabeth and James Rivers. *Author's collection*

The U.S.S. *Merrimack* during her glory days as an auxiliary steam frigate in the United States Navy. Even during the best of times her engines were notorious for unreliability and consumption of coal. *United States Naval Historical Foundation photo*

Destruction of the Gosport Naval Shipyard on April 20, 1861. The top image shows the massive shiphouses as they became fully involved. The lower image shows the U.S.S. *Pawnee,* her deck crowded with soldiers, and

the tug *Yankee* towing the sailing vessel U.S.S. *Cumberland* to safety. Behind *Cumberland*, the *Merrimack* and the shiphouses are consumed by flame. *United States Naval Historical Foundation photo*

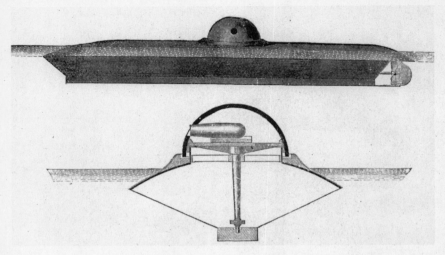

John Ericsson's original concept for an ironclad steam battery with a revolving turret. These plans were submitted to Napoleon III in 1854, but the French took a pass on the idea. *From Ericsson,* Battles and Leaders

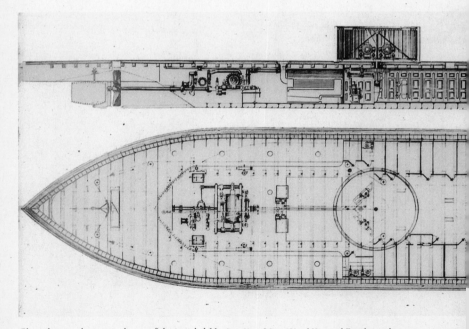

Plans showing the interior layout of the ironclad *Monitor. United States Naval Historical Foundation photo*

John Ericsson, the mercurial genius from Sweden who designed the *Monitor*. The photo suggests the great physical strength for which he was known. *United States Naval Historical Foundation photo*

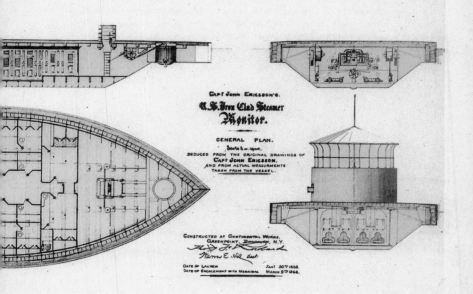

Artist's rendition of the final stages of *Merrimack's* transformation into the ironclad *Virginia* in the dry dock at the Gosport Naval Shipyard. There are no known photographs of C.S.S. *Virginia*. *United States Naval Historical Foundation photo*

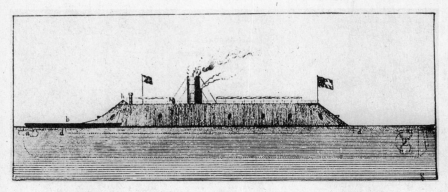

A profile drawing of *Virginia*. Though many of the numerous drawings of the ironclad are highly imaginative, this one appears to be quite accurate.
United States Naval Historical Foundation photo

LEFT: Franklin Buchanan, C.S.S. *Virginia's* first commanding officer. *United States Naval Historical Foundation photo*

BELOW: Commander Catesby ap R. Jones. Jones had hoped to get command of *Virginia*, but he was thought too young and inexperienced for the position. As first officer, he assumed command during the ship's most memorable day of fighting. *United States Naval Historical Foundation photo*

ABOVE: "Old Buck" and "Old Tat,"
the C.S.S. *Virginia's* only two
commanding officers, Franklin
Buchanan and Josiah Tattnall.
From Wood, Battles and Leaders

RIGHT: John L. Worden, first
and most famous captain of the
U.S.S. *Monitor. United States
Naval Historical Foundation photo*

The officers of the U.S.S. *Monitor* in a relaxed pose. Front row, left to right: Third Assistant Engineer Robinson Hands, Second Assistant Engineer Albert Campbell, and Acting Master Edwin Gager. Middle row, left to right: Acting Master Louis Stodder, Paymaster William Keeler, Lieutenant William Flye, and Surgeon Daniel Logue. Back row, left to right: Acting Master's Mate George Fredrickson, Third Assistant Engineer Mark Sunstrom, Lieutenant Samuel Dana Greene, Lieutenant L. Howard Newman from the *Galena* and Engineer Isaac Newton. *United States Naval Historical Foundation photo*

U.S.S. *Monitor* on the James River. The officer grasping his lapels is Acting Master Louis Stodder. To the left of Stodder is Robinson W. Hands. Seated is Albert Campbell, and the man with the straw hat is Lieutenant William Flye. Note the visible dents in the turret from *Virginia's* guns and Stimers's new and improved pilothouse with sloping sides. *United States Naval Historical Foundation photo*

The *Monitor* Boys: the crew of the ironclad cook on deck during the dull, sweltering summer the ship spent on the James River. *United States Naval Historical Foundation photo*

U.S.S. *Cumberland* is rammed by C.S.S. *Virginia*. With the ironclad wedged in her side, *Cumberland* nearly took Virginia to the bottom. *From Wood,* Battles and Leaders

The defiant men of the *Cumberland* fight to the last, firing even as their vessel sinks under them. *Cumberland's* gunners managed to inflict the only real damage that *Virginia* would suffer in two days of fighting. *United States Naval Historical Foundation photo*

In a vain attempt to flee the *Virginia*, *Congress* sets her fore topsail and heads for shore. Remaining in deeper water, *Virginia* fires into her stern, while off in the distance on her starboard side, ships of the James River squadron add their firepower. *From Wood,* Battles and Leaders

LEFT: Hard aground and on fire, with no guns that would bear on the enemy, *Congress* is abandoned by her men as the first officer surrenders the ship. *From Wood,* Battles and Leaders

BELOW: *Monitor* and *Virginia* pounded one another at extremely close range for four hours, but neither was able to inflict serious damage on the other. Note in this quite accurate engraving *Virginia's* submerged after deck, just visible beneath the water. *From Wood,* Battles and Leaders

A fairly accurate depiction of the first battle between ironclads. Note *Virginia's* stack, riddled with holes, which did little to help her already poor draft. Spent shot lies on *Monitor's* deck, waiting to become souvenirs. Ericsson's famous anchor well is visible forward of the pilothouse. *United States Naval Historical Foundation photo*

Boats from the U.S.S. *Rhode Island* brave the seas to take men off the foundering *Monitor*. While the depiction of *Monitor* is not terribly accurate, the artist has correctly included the lifelines around the deck that saved many lives that night, as well as the boilerplate shield added to the top of the turret for protection against small arms fire.
United States Naval Historical Foundation photo

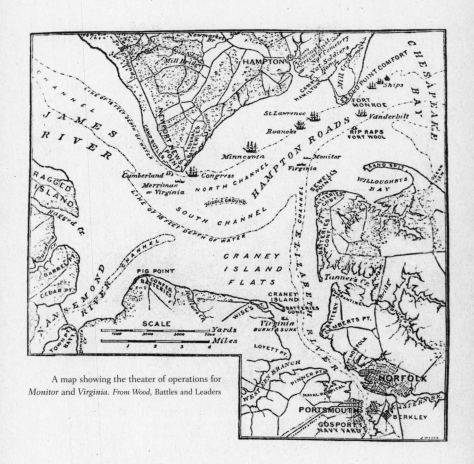

A map showing the theater of operations for *Monitor* and *Virginia*. *From Wood, Battles and Leaders*

The dry dock in which *Merrimack* became *Virginia* as it appears today. The Gosport Naval Shipyard, now known as the Norfolk Naval Shipyard, is still one of the navy's most important facilities, and has been vastly expanded since the days of the Civil War.
Virginia's dry dock is still in use. *Author photo*

ity." An *ad hoc* gunnery range was set up at the naval hospital near the Gosport Naval Shipyard. Jones was allowed to requisition gunpowder for the tests, but no more than 300 pounds.

The lack of gunpowder would be a constant impediment to training and experimentation in the Confederate navy. The Inspector of Ordnance, A. B. Fairfax, complained, "we have been deterred from any experimental firing with rifle guns since August, till authorized by the Department to try the *Merrimack* guns and projectiles, by the scarcity of powder."

Brooke produced two different calibers for *Virginia*. The pivot guns at the bow and stern were 7-inch rifles, weighing 14,500 pounds each. The other two were 6.4-inch rifles placed on the broadsides. These weighed 9,000 pounds each. Both calibers had 3-inch steel bands shrunk around the breach. The tests that Jones performed proved the guns to be a great success. The 7-inch gun, with a 12-pound charge, hurled a 100-pound shell 4½ miles. At the close range at which an ironclad was intended to fight, that punch would be devastating.

Tredegar cast the four Brooke rifles. Six 9-inch Dahlgren smoothbore shell guns, courtesy of the U.S. navy, filled out the ironclad's broadsides.

Along with the guns, Tredegar was manufacturing shell, and again *Virginia* took precedent. In reply to a request for ammunition, Commander John R. Tucker, captain of the CSS *Patrick Henry* on the James River, was told, "The shells are not yet ready, but will be forwarded as soon as the orders for the *Merrimack* are filled." Because of the delays with manufacturing and shipping, Jones suggested that shells be sent as they were made, in lots of fifty at a time.

The Brooke rifle was designed to fire shells as well as solid shot, or "bolts." Given the time constraints, Tredegar was able to make one or the other, but not both. Shells fired from rifled guns were the hot new trend in ordnance, so they were the focus. Robert Minor wrote to Jones, "As rifled guns and projectiles are all the go, I am kept hard at work preparing them for service."

More to the point, shells were the most effective ordnance against the wooden ships *Virginia* was expected to meet. "[O]wing to the fact that the enemy had no ironclad afloat at the time she [*Virginia*] first went out," Brooke testified, "and there being a great pressure upon the works for projectiles of other kinds proper to use against wooden vessels, she was not furnished with bolts."

The only solid shot aboard the ship was for the 9-inch Dahlgrens and there were just a few, intended to be used as heated shot. They were cast a bit smaller than normal shot to compensate for the metal's expandsion when heated. A system was devised by which the shot was heated in the two forward furnaces in the boilers, run up to the gun deck in the ash hoist, and loaded into the nearest Dahlgrens. The system was "found to work with great celerity."

Brooke's tests indicated that a solid bolt fired from his 7-inch gun would have pierced *Monitor*'s 8-inch-thick turret, but no such shot was available. Had *Virginia* been provided with solid shot, the Battle of Hampton Roads might have been a very different affair.

As the iron plating was bolted to the shield, and the guns tested, John Luke Porter installed another device of his own design—an iron ram. "Very little was known about them at this time," Porter wrote. Certainly anyone with any significant experience fighting with a ram had been dead for thousands of years. Once sail replaced the oar as a man-of-war's motive power, the ram was no longer operational. The advent of the steam engine made it a practical weapon once again.

While Brooke had felt that the ship's reinforced cutwater would be sufficiently strong to use the bow as a weapon, Porter opted for an actual ram that projected beyond the cutwater. Porter made his ram out of cast iron, "for the want of something better." It was wedge-shaped, weighed 1,500 pounds, and was bolted to the bow. It probably stuck out around 2 feet beyond the cutwater, though contemporary descriptions are vague.

Ironclad Sailors

As 1861 gave way to 1862, the pace of rebuilding *Merrimack* continued as frenetically as ever. More so, perhaps, as the Ericsson Battery was now hanging over the heads of the Confederate navy, and the iron from Richmond, the want of which had caused such great delays, was now starting to flow into Gosport.

As the ship's launch date approached, Mallory and Franklin

Buchanan, head of the Office of Order and Detail, began to think about manning. By January, most of the officers had been appointed to the ship. Besides Catesby Jones, the executive officer, the other lieutenants were Robert D. Minor, who served as flag lieutenant, Charles Simms, Hunter Davidson, John Taylor Wood, J. R. Eggleston, and Walter Butt.

It was less clear who would be chief engineer. H. Ashton Ramsay had been filling the slot since Williamson's departure, but Ramsay's commission as chief engineer had not yet been approved, and without it he could not hold the position. Initially it was thought that William Williamson would take the position, but for some reason the order was revoked, as were the orders for two other chief engineers. While all of that dithering was going on, Ramsay's acting chief engineer's commission came through, and he was given the job he had already been doing for months.

Ramsay's appointment caused problems when it came to recruiting the black gang. At twenty-five years old and as a newly minted chief engineer, Ramsay was junior to nearly every other engineer who had left the old navy to come South.

A ship the size of *Virginia* would normally have carried, along with the chief engineer, two first assistants, two second assistants, and four third assistant engineers. Ramsay was unlikely to find that many qualified men, nor did he necessarily need them. An engineering department of that size was meant for a ship that would be steaming for at least twenty-four hours at a time. Most people associated with *Virginia* understood that she would not be going far enough to even require a change of watch.

In the end, Ramsay settled for one first assistant who had been a second assistant in the U.S. navy, and two civilian third assistants who, though they had experience with steam engines, had never been to sea.

If engineers were hard to come by, sailors were proving nearly impossible to find. It was a problem that would always plague the Confederate navy. With no merchant marine to speak of, the South did not have a core of experienced mariners from which to draw. "In the old service," Lieutenant John Taylor Wood explained, "the majority of officers were from the South, and all of the seamen from the north."

To make things worse, many experienced sailors had joined the army at the outbreak of the war, just as Ashton Ramsay had done. With manpower deficiencies of their own, the army was very reluctant to turn men over to the navy.

French Forrest tried to entice men into the naval service. In late November, more than two and a half months before the ship was even launched, he wrote, "I have reconsidered the matter in regard to the bounty to be given to the men who are to be detailed for the *Merrimack*." A bounty of twenty dollars would be given to seamen and ten dollars to ordinary seamen who were discharged from the army and enlisted in the navy for two years.

In mid-January he reconsidered the matter again and allowed men to sign on for as little as six months, though bounties would be paid only to those who joined for three years.

In early February, Buchanan expressed his frustration with the situation when he wrote, "I have a rendezvous open in Richmond and Norfolk to ship men for the Navy. As yet but few have offered. The *Merrimack* has not yet received her crew, notwithstanding all my efforts to procure them from the Army."

By the end of February time was running out. Forrest authorized Paymaster Richie to give a bonus of fifty dollars to seamen of any description and landsmen as well (except boys) "who may ship for three years, or the war, aboard the *Virginia*" (it had, by then, been commissioned).

A few sailors were found in Norfolk, refugees from the gunboat flotilla of the North Carolina inlets and its disastrous attempt to hold back the Union invasion. In fact, a number of the crew first shipped onboard *Virginia* had been sent off to temporarily join the flotilla rather than sit idle in the navy yard (including a gunner whom Forrest called "Double Block Jack, who comes recommended by Captain Buchanan"). These men were repatriated with their ship, but more were needed.

For more, Lieutenant Wood was sent to see General John Magruder, in command of army forces at Yorktown, who was rumored to have two battalions from New Orleans that might have among their ranks some seasoned mariners. Wood was taken to each of the camps, where the men were paraded to hear the lieutenant explain what he was after. Around two hundred men stepped forward, of whom Wood chose eighty who were experienced sailors or gunners. The same routine was repeated in camps around Richmond and Petersburg.

Still the army tried to thwart Wood's efforts, and dispose of their riffraff as well. Arriving back in Norfolk from one recruiting trip, he found among the men who had been sent only two that he had picked, the rest "a very different class of men from those I selected." He complained to Buchanan, "We can not get good men from the Army unless their officers will assist and not oppose us."

In the end, a crew of 320 men were assembled, all volunteers, some quite willing. Captain R. C. Taylor of the 6th Reserve Volunteers on Craney Island offered his entire company to man the ironclad's guns.

The men who joined the *Virginia*'s crew were willing and brave, but they were not, for the most part, men-of-wars men, save for a "sprinkling . . . whose value at the time could not be overestimated." Most had never even seen guns such as those they would be firing in battle.

With the ship still under construction, the men could not live or drill onboard *Virginia*. Instead, the officers used the old frigate *United States**, part of the federal largess that came with the abandoned yard. The *United States* served as the receiving ship for the *Virginia*'s crew, a place for them to live while their new ship was prepared, as well as a "school ship." Every day for weeks the men drilled at the obsolete guns onboard the outmoded frigate. The first time they would fire the *Virginia*'s guns was when they were pointed at an enemy ship.

Ramsay had to draw his firemen and coal passers from the men who had volunteered. "I selected my firemen with great care, insisting on receiving none but strong, able-bodied men." He managed to find thirty who met his standards, and they would keep *Virginia*'s four voracious boilers fed with coal as she steamed into battle.

"Mr. Porter, Do You Really Think She Will Float?"

In the second week of February, the *Virginia* was ready for a test float in the dry dock. This was not to be her official launching; it was just a

---

*In official Confederate records, this ship was generally called the "U.S." or the "States." The name *United States* was apparently either too long or too repugnant to write.

test to see if she would indeed bear the weight of the shield, or if she would turn turtle as so many thought.

John Luke Porter was anxious, indeed he had been anxious for a while. "I received but little encouragement from anyone while the *Virginia* was progressing," Porter wrote. "Hundreds—I may say thousands—asserted she would never float. Some said she would turn bottom-side up . . . public opinion generally around here said she would never come out of the dock. You have no idea what I suffered in mind since I commenced her."

Porter was subject to the same kind of ferocious scepticism that Ericsson suffered, and was getting the sort of encouragement from his superiors that Ericsson was getting from Commodore Smith. When Porter reported the *Virginia* ready for a float test to the yard's executive officer, Sidney S. Lee, Lee replied, "Mr. Porter, do you really think she will float?"

But Porter was not able to dismiss such skepticism as easily as the self-assured Swede, perhaps because Porter had nothing like Ericsson's background in engineering and design.

After *Virginia*'s success, Porter would make much of the fact that he alone had been responsible for calculating her center of gravity and the amount of weight she could carry. It was that same accountability that weighed on him as the water rose up in the dry dock and embraced the ship's copper-clad bottom.

Slowly she lifted up off the chocks, the old *Merrimack*'s hull waterborne again for the first time in nine months. She floated. She did not roll over. But she floated high out of the water, considerably too high.

Somewhere in Porter's calculations, which were admittedly very difficult given the circumstances, he had made an error. *Virginia* floated so high that her wooden sides were exposed beneath the iron plate. She was as vulnerable as if she had no iron shield at all.

Chapter 22

Monitor—person or device for checking or warning.

—THE OXFORD DICTIONARY
OF CURRENT ENGLISH USAGE

On January 16, the United States' first ironclad warship was commissioned. In fact, seven of them were commissioned. They were the "Pook Turtles," *Cairo, Carondelet, Cincinnati, Louisville, Mound City, Pittsburg,* and *St. Louis.* They were each 512 tons and mounted around a dozen guns in broadsides and through the front and back of their iron casemates. They were propelled by centerline paddle wheels, ideal for river work. The turtles were formidable machines, the likes of which the Confederates could not hope to build.

In New York and Norfolk, the arms race continued.

With the *Monitor* construction well under way, Commander David Dixon Porter was asked by the Navy Department to inspect the vessel and report on its viability. Porter, son of the famed War of 1812 commodore David Porter, was already something of a golden boy in the navy. He would end the war as a rear admiral.

Porter was pleased to undertake the assignment and left Washington for New York City. Most likely, Porter knew Ericsson's reputation. In any event, he decided to have some fun by tweaking the hot-tempered engineer.

Porter arrived at Ericsson's office at 95 Franklin Street and showed Ericsson his orders. "Well, you are no doubt a great mathematician,"

Ericsson said to Porter, "and know all about the calculations which enter into the construction of my vessel."

The mere fact that the navy had sent yet another officer to evaluate the battery must have irritated Ericsson, but Porter managed to make it worse. He admitted he was not a mathematician. "But I am a practical man, and think I can ascertain whether or not the *Monitor* will do what is promised for her."

"Ah, yes!" Ericsson replied. "A practical man! Well, I've had a dozen of those fellows here already, and they went away as wise as they came. I don't want *practical* men sent here, sir. I want men who understand the higher mathematics that are used in the construction of my vessel."

Ericsson was not mollified when Porter assured him that he knew "the rule of three, and that twice two are four." In fact, as Porter reported it, Ericsson's "hair bristled up, and the muscles of his brawny arms seemed to swell as if in expectation of having to eject me from the room."

Porter suggested that Ericsson could give him instruction in higher mathematics, something Ericsson declined to do. Ericsson did, however, allow Porter to look over the plans that he had drawn up. "Look at this drawing and tell me what it represents," Ericsson demanded.

"It looks like a coffee mill," Porter replied.

"On my word of honor, young man, you are vexing and I am a fool to waste my time on you." But Porter had Ericsson hooked and continued to play him. Ericsson handed Porter a model of the steam battery and asked what he thought.

Holding the model upside down, Porter ran his hand over the bottom (now on top) and said, "This is evidently the casemate," and pointing to the turret, now on the bottom, said, "and this is undoubtedly where you carry the engine."

Ericsson was not one to recognize humor where his inventions were concerned. Porter continued his bull baiting for a while longer, then suggested they go and examine "the Simon-pure article." The two men took the ferry across the East River to Greenpoint and the Continental Ironworks. Together they spent an hour climbing all over the vessel, examining every part of it. Porter continued to toy with Ericsson the whole time.

When they were done, Porter said, "Now, Mr. Ericsson, I have borne a good deal from you today; you have mocked at my authority and have failed to treat me with the sweetness I had a right to expect. I am about to have satisfaction, for on my report depends whether or not your vessel is accepted by the department, so I will tell you in plain terms what I think of your 'iron pot.'"

Ericsson was unimpressed. "Say what you please, nobody will mind what you say."

But to Ericsson's surprise, Porter told him that his report would claim, "Mr. Ericsson has constructed the most remarkable vessel the world has ever seen—one that, if properly handled, can destroy any ship now afloat, and whip a dozen wooden ships together."

It was not what Ericsson had expected, but he was delighted. He grabbed Porter's hand and shook it, exclaiming, "All this time I took you for a damned fool, and you are not a damned fool after all!"

According to Porter, the two men were the best of friends from that moment on.

Alban C. Stimers

By November 15, twenty-one days after the keel laying, the navy's building superintendent certified that the first fifty thousand dollars' worth of work had been finished. The middle section of the lower vessel was mostly done and the deck beams were being fitted. Ericsson estimated that the following week the engines and part of the machinery to run the turret would be in place.

The superintendent doing the certification was Chief Engineer and Constructor Alban C. Stimers, a twelve-year veteran of the U.S. navy, former chief engineer of the USS *Roanoke*, and the last chief engineer to serve aboard *Merrimack* before she was scuttled and burned. Stimers was thirty-four years old, with wavy hair, a thick mustache that trailed into bushy sideburns, and a John Ericsson-like stubbornness and temper.

Not surprisingly, Ericsson felt that the most important officer aboard his battery would be the chief engineer, and he felt that Stimers, whom he already knew and respected, was the man for that position. On October 4 he wrote to Smith, saying, "As an engineer of the highest intelligence will be required for the battery, the success mainly depending on that officer as the whole is machinery, permit me to suggest Chief Engineer A. C. Stimers of the *Roanoke*."

In the U.S. navy of the Civil War, engineers were classified as chief engineer or first, second, or third assistant engineer. Chief engineers were commissioned officers, like lieutenants and captains, while the assistants were warrant officers, like surgeons or paymasters. As a chief engineer, Stimers could not be appointed to a vessel as small as the Ericsson Battery. However, on November 5, Gideon Welles directed that Stimers be appointed "Superintendent of the iron-clad vessel or Battery." As such, Stimers would oversee the construction of the vessel, looking out for the navy's interests, and determine when sufficient work had been completed for the release of payments.

Had the superintendent's position gone to a man with whom Ericsson butted heads, the result might have been a disaster. Fortunately, Stimers and Ericsson were cut from the same cloth. Their personalities and vision for the floating battery were well matched. Indicative of their close relationship is the fact that all of Stimers's official correspondence was addressed to Ericsson's home at Franklin Street.

Commodore Smith did not let up in his relentless gloom and doom. By mid-October he informed Ericsson that he had requested guns for the battery from the Ordnance Department, but the Ordnance Department informed him that the guns would never be used, presumably because the vessel would never float. A few days later, he added, "I have nothing more to say on the subject but that the Government will fall back on the contract in case of failure." Ericsson would have been lucky if Smith had nothing more to say, but as it happened, the old man still had a lot of worrying left in him.

"Hard Knocks"

Smith was not alone in worrying. Fear of *Merrimack* was already well entrenched among the naval officers in Washington and Hampton Roads. By the time construction was even begun on the Ericsson battery, nearly half a year had passed since the loss of the Gosport Naval

Shipyard and the subsequent refloating of the scuttled and burned steam frigate. Intelligence trickling in from Norfolk fueled the paranoia of the naval officers in Hampton Roads.

On the very day that the contract for *Monitor* was signed, Flag Officer Louis M. Goldsborough, commanding the Atlantic Blockading Squadron, wrote to Welles from Hampton Roads, "[B]y intelligence received to-day, the enemy is about to make a demonstration on Newport News with his vessels at Norfolk, so as to get his steamers down the James River at all hazards. My information also says that he is determined to try the *Merrimack,* with her iron casing, against this place."

Two weeks later, Goldsborough described the plans he was making to counter *Merrimack* when she came out. His idea was to lure the ironclad to a place where she could be caught between the broadsides of *Congress* and *Cumberland* on the one hand, and *Roanoke* and whatever other ships were available on the other. "Nothing, I think, but very close work can possibly be of service in accomplishing the destruction of the *Merrimack,* and even of that a great deal may be necessary."

Neither Goldsborough nor anyone else in the North could yet appreciate the near-absolute invulnerability the ironclad would possess, and indeed they never would, until that first broadside rattled off her casemate.

William Smith, commanding the *Congress,* was less confident about meeting the ironclad. After describing to Goldsborough the elaborate precautions he had taken to shield his vessel from "fire craft, torpedoes, or infernal machines," he told the admiral, "I have not yet devised any plan to defend us against the *Merrimack,* unless it be with hard knocks."

Goldsborough's intelligence informed him that *Merrimack* still needed quite a bit of iron to complete her casemate. Regardless, he felt that the ship would be out within two weeks (which was understandable, considering that French Forrest, in command of the shipyard where she was building, thought so as well). His anxiety permeated the Navy Department, which understood that the real best hope was for the Ericsson Battery to blow the whole works away while she was still in the dry dock. Letters from Smith continued to arrive at the Franklin Street address.

On December 5, Smith wrote in a panic to Ericsson. The commodore had been informed that the turret would not be ready to leave the shop for another month. "I beg of you to push up the work," he urged. By way of encouragement he added, "I shall demand heavy forfeiture for delay over the stipulated time of completion. You have only thirty-nine days left."

Ericsson *was* pushing hard. By December 17, Stimers had certified the completion of $150,000 worth of work, though they had yet to receive even the first payment from the government, requiring Winslow to borrow the money from his bank in order that construction might continue. In fact, one of the reasons that *Monitor* was not delivered within the one hundred-day window was the navy's want of diligence in issuing payments.

At the Novelty Ironworks, the turret was coming together piece by piece. With no possibility of moving the entire structure, the turret was built so it could be disassembled and moved in parts to Continental Ironworks, where it was reassembled on the hull.

To accomplish this, the first layer of 1-inch-thick iron was bent to the proper curve, punched with holes for the bolts, and fitted to a wooden framework built to the turret's inside dimensions. The second layer of iron was then bent and placed over the first, with each piece overlapping the joints of the first layer. A dowel dipped in paint was thrust through the bolt holes of the inner layer to mark the bolt positions on the second layer, which was then taken down and punched. In that manner each of the eight layers of iron was shaped and punched. Finally the whole thing was reassembled on the framework and the bolt holes were reamed out to eradicate any misalignment.

Each gun port was made up of three overlapping holes cut in a vertical line through the iron plates of the turret. The points where the arcs of the circles intersected were to be smoothed out, but in the rush of construction this was not done. Photographs of the *Monitor* clearly show the untrimmed edges.

By mid-December, all of the machinery was onboard the vessel and the deck plates and the armor for the sides were rolled and delivered to the Continental Ironworks. Ericsson estimated that within two weeks steam would be raised and the machinery set in motion.

Around that time, Gideon Welles received a detailed and accurate report on the state of affairs at Gosport. The informer was one W. H. Lyons, a Union man who had applied for the job of master machinist at the navy yard shortly before it was taken over by the Confederacy.

Lyons reported, "The steamer *Merrimack* is now in the dry dock. She has been cut down 2 feet below the water line and a roof or covering of wood 28 inches thick at an angle of 30 degrees. There is 10 feet of the top flat, with a heavy grating of wrought iron, and the balance is covered with plates of iron 3 inches thick. The bow has a casting made in the shape of wedge and weighing 2 tons for the purpose [of] sinking ships."

In early January, Ericsson wrote to Smith pointing out that though they had completed $200,000 worth of work, they had received from the government only $37,500 (one payment of $50,000 less the 25% the government held in reserve). "[M]y contemplated organization and operation of what is called night gangs has been to some extent frustrated." As eager as the government was to have the battery done, its lethargy in issuing payments was not helping.

Down the Ways

On January 30, 1862, around ten o'clock on a cold, rainy morning, the *Monitor* slid down the ways and into the icy East River. Standing on her deck were John Ericsson and several of his associates, as well as some of the officers already appointed to the ship. Standing around in the muddy shipyard, crowds of onlookers watched the odd vessel's launch. Many had come with the expectation of seeing her hit the water and keep on going straight to the bottom. More than a few had placed bets that she would do just that.

The chief cause of concern was the overhanging upper vessel. Because the upper vessel extended so far out over the lower vessel, a good deal of it would be in the water before the buoyant lower vessel was even off the ways. Many experienced observers felt that the ship would continue on its downward trajectory until it hit the bottom of the river. One pundit quipped, "If Ericsson ever finds his battery after she is launched, he will have to fish her up from the mud, into which her stern will surely plunge."

Thomas Rowland of the Continental Ironworks also had his doubts. More than an onlooker, Rowland was responsible for the launch, his contract stating that the Continental Ironworks would launch the battery at its own "risk and cost." To counteract this potential problem, Rowland constructed two wooden tanks that he chained to the underside of the upper vessel, all the way aft. The battery hit the water and floated free, the upper vessel lifted by the air-filled tanks. Once it was clear that she would float on her own, valves on the tanks were opened, the tanks flooded, and the ship settled down onto her waterline. Fifteen minutes later she was floating on an even keel.

The people crowding the yard, civilians, navy men, and yard workers, cheered, shouted, and waved their hats, even those who had just lost money on the floating battery. One reporter from the *New York World* wrote it was "evident even to the dullest observer that the battery hadn't the slightest intention of sinking."

Monitor was supposed to be built in one hundred days. The *Monitor* builders, their supporters, and historians have employed various formulas to spin the ship's construction time in such a way as to claim she was. Ericsson himself would state that the ship was launched one hundred days from the laying of the keel plates. That was true, but it was not what was promised. The contract called for completion one hundred days from the date of the contract, not the keel laying. As Commodore Smith's panicked missive correctly pointed out, January 12 was the one-hundred-day mark. January 30 was 118 days after the contract was signed.

Other historians, and perhaps some of the *Monitor*'s builders, counted working days only. Assuming a six-day work week, then the vessel was launched 105 days after the contract was signed.

All of those calculations are irrelevant, however, when one considers that the contract did not call for the vessel to be launched in one hundred days, but rather that in one hundred days it "in all aspects shall be completed and ready for sea." On January 30, it was certainly not that.

Welles, for all his admiration of the ship and satisfaction with her builders, was quick to point out that she had not been completed in time. He dated her completion at March 3, when she was turned over to the government, forty days after she was due. In hindsight it was clear that had *Monitor* been ready for sea on January 12, she might well have destroyed the *Virginia* in dry dock and spared the Union the horrible carnage of *Congress* and *Cumberland*. Welles never lost sight of that fact.

Prior to the launch of the floating battery, Gustavus Fox wrote to Ericsson asking him what he proposed to name the ship. Ericsson was drawn to a word much more in common use in the mid-nineteenth century than it is today. He saw his novel man-of-war as a teacher of lessons. Impervious to shot, it would teach the leaders of the Confederacy that they could no longer deny entrance to Southern waterways with batteries onshore.

But Ericsson had others in mind as well. With tension mounting between the United States and Great Britain over the removal of Mason and Slidell from the *Trent*, Ericsson was looking at his rivals across the ocean. And no doubt he recalled the pointed indifference with which the Royal Navy had greeted his screw-driven steamer.

"Downing Street" will hardly view with indifference this last "Yankee notion," this monitor. To the Lords of the Admiralty this new craft will be a monitor suggesting doubts as to the propriety of completing those four steel-clad ships at three-and-a-half millions apiece. On these and similar grounds I propose to name the new battery Monitor.

Monitor she would be.

Chapter 23

"The Vessel Was Called . . . *Virginia*."

John Luke Porter had miscalculated.

In his original estimates of *Virginia*'s buoyancy, he had figured on an extra 50 tons of displacement, meaning that he would have to add 50 tons of ballast to bring her down to the correct waterline, with the eaves of the shield two feet below the surface.

Raising the waterline to avoid cutting into the propeller left him instead with an extra 200 tons displacement. That was the reason Porter could assure Brooke that the ship could bear four inches of plate as opposed to three. Porter knew that the extra iron alone would not put the eaves 2 feet under water. He further compensated for the surfeit of buoyancy "by putting that amount of kentledge [pig iron used as permanent ballast] on her ends and in her spirit room in order to bring her eves down two feet." Despite sacrificing liquor storage for pig iron, it was still not enough.

John Mercer Brooke was not happy about *Virginia* floating so high out of the water. Nor was Catesby Jones. Brooke would later testify that "Mr. Porter stated to me that he had accidentally omitted in his calculation some weights which were on board the ship, in consequence of which she did not draw as much water when launched as he anticipated." The weights Porter forgot about, apparently, were the

Merrimack's masts, rigging, and spars, which, of course, were no longer there. If Porter did admit that, it was before he and Brooke began their protracted fight over credit for the ship's design. Later, Porter would never own up to having made a mistake.

Floating in the dry dock for the first time, *Virginia* drew about 19 feet. Getting her eaves submerged would bring her down to about 21 feet of draft forward and 22 aft. This would be a major hindrance in the shallow waters in which she would operate, worse than Brooke, Porter, or Williamson had anticipated, but there was nothing for it. The shield had to be submerged.

Finally, on February 17, *Virginia* was ready for launch. It was around this time that the navy men began to call the ship *Virginia*. Two days before the launch, Commodore Forrest referred to her as the "*Merrimack*" (he had on occasion called her "CSS *Merrimack*"). The day before the launching he called her the "*Virginia* (late *Merrimack*)." Prior to that date she does not appear to have ever been called *Virginia*. After that date she was almost never called *Merrimack*, at least in all official Confederate reports and most private correspondence.

For all the enormous significance of *Virginia*, and the terrific effort that went into completing her, the launch was surprisingly anticlimactic. There were only five men aboard her, four marines and a corporal. There was none of the pomp and ceremony that traditionally attends a launching and christening. As one of the few witnesses put it, "no sponsor nor maid of honor, no bottle of wine, no brass band, no blowing of steam whistles, no great crowds to witness this memorable event."

With French Forrest clamping down on access to the yard, there were none but navy men and officers there. Of the officers, only one, Marine captain Reuben Thorn, was part of *Virginia*'s crew. The rest of *Virginia*'s men were onboard the former *United States*, preparing for the moment *Virginia* would steam into action.

Water flowed into the granite dry dock, and the ironclad once again lifted off its blocking and floated free. This time the massive dry dock gates were open and the ship was warped out into the Elizabeth River. More than eight months after she was begun, and eighteen days after *Monitor* was launched, the former USS *Merrimack* was christened the Confederate States Ship *Virginia*. But she was still far from complete.

Once she was in the river, *Virginia* was brought around to the huge shears that the Federals had cut down and the Confederates reerected. The sheers were used to lift the ironclad's ten guns and set them down through the shield deck onto the gundeck below.

The total weight of the ordnance was more than 51 tons, but still *Virginia* was floating too high in the water. Porter now had to estimate how much farther she would sink under the weight of the coal, powder, shot, and stores that would be brought aboard, and how much she would subsequently rise as those materials were expended. By his reckoning she still needed more ballast. He loaded the bow and stern with "blocks of iron kentledge laid loosely on the decks."

Despite all the work that remained, and the fact that the ship had no coal, powder, shot, or captain, French Forrest seemed to feel her departure was imminent. As far back as early December, the optimistic flag officer had ordered obstructions removed from the river, saying, "The period is fast approaching when it is presumed that the *Merrimack* will be in readiness to proceed down to Hampton Roads."

Five weeks later she was still in dry dock when Forrest wrote to General Huger, commanding the army in Norfolk, "There can be little question that communications are held by the unfaithful people of these two towns with the enemy, and the successful undocking of the *Merrimack* will, without doubt, be communicated by means of signal rockets or lights."

Forrest suggested that all signal officers be provided with rockets, so that when the "unfaithful people" sent up their signal, the faithful could signal as well, making it appear as if they were all signaling one another. Apparently no signals of that kind were ever made.

Forrest also suggested to Huger that "in view of the approaching departure of the *Virginia*" all communications with Fortress Monroe by flag of truce be suspended.

On the day *Virginia* was launched, Forrest wrote to Catesby Jones, saying,

> SIR: *You will be pleased to receive on board the* Virginia, *immediately after dinner today, all the officers and men attached to the vessel, with their baggage, hammocks, etc., and have the ship put in*

*order throughout. She will remain where she is to coal and receive
her powder.*

*You will report to me when your men and officers are on board,
and use every effort to get the ship in order, as this day she is put in
commission.*

Moving onboard *Virginia* must have made the men long for the decrepit old *United States.* Jones recalled the ironclad "was . . . crowded
with mechanics until the eve of her fight. She was badly ventilated, very
uncomfortable and very unhealthy. There was an average of fifty or
sixty at the hospital, in addition to the sick list on board." Since her
shield deck was only a grating, there was nothing to keep the rain off
the gun deck, and the gun deck leaked onto the berthing deck below.
There was almost no dry place on the ship, which must have con-
tributed mightily to the men's ill health.

Old Buck

On February 24, Mallory named Franklin Buchanan flag officer of the
James River squadron, which did not surprise many. Rumor had been
circulating for five months that "Old Buck" would have command of
the ironclad. Indeed, weeks before the appointment, Buchanan was already making the kind of decisions a commanding officer would make.
While *Virginia* was still in dry dock, Forrest told John Luke Porter,
"Captain Buchanan is anxious to carry two small, light boats, hung at
davits, with him. I have no objection to this."

Despite the appointment, Buchanan was not, in fact, *Virginia*'s captain. Strictly speaking, she never had a captain.

Mallory wanted Franklin Buchanan, but there were a number of
captains senior to him, including French Forrest, who wanted command of *Virginia.* Since his days in the U.S. Senate, Mallory had little
patience for promotion due to seniority. To avoid the whole issue, and
to get the man he wanted, he made Buchanan a flag officer of the
squadron. As such, he was allowed to choose in which vessel he would
hoist his flag. Unsurprisingly, he chose *Virginia.*

It was well worth any subterfuge required to get Franklin Buchanan in command of the *Virginia*. Uncompromising, bold, driven, Buchanan was exactly the man for that ship.

Franklin Buchanan was born on September 11, 1800, in Baltimore, Maryland, the son of a successful doctor and grandson of a signer of the Declaration of Independence. At age five he moved with his family to Pennsylvania, where his father had been appointed, through family influence, as attending surgeon at the Lazoretto Hospital. Two years after that, Buchanan's father died, leaving behind Franklin and his six siblings.

In 1815, at the age of fifteen, Buchanan was appointed a midshipman in the U.S. navy, just missing one of the United States' great naval conflicts. The War of 1812 was in its waning days. Indeed, a peace treaty had already been worked out in England, but no one in the United States yet knew it.

Buchanan first learned the ropes aboard the frigate *Java* under the famed Captain Oliver Hazard Perry. Their station was the Mediterranean, where the United States was still grappling with the Barbary pirates. Two years later he found himself in the midshipman's berth aboard the 74-gun USS *Franklin*. Among his shipmates were midshipmen Samuel Francis Du Pont and David Glasgow Farragut.

Franklin Buchanan's years as a midshipman were spent primarily in the Caribbean, fighting pirates from the deck of small, fast ships. It was good duty for an ambitious and restless young officer, with opportunities to exert independent command that would not be available on a frigate or a ship of the line. In 1825, after ten years in the service, Buchanan was promoted to lieutenant.

For the next sixteen years Buchanan served primarily in the Mediterranean, aboard such famous vessels as *Constellation* and *Constitution*. During that time, he developed into a first-rate seaman and naval officer, a man devoted to his duty and to strict adherence to rules and regulations. Indeed, young Buchanan developed a reputation, not always welcome, of being a stern disciplinarian who was not afraid to use floggings to enforce authority. He had no tolerance for mediocrity, and tended to see things in black-and-white terms, viewing officers and men as either worthy or worthless. It was a characteristic that was as integral a part of him as his prominent nose.

Buchanan was given his first command, the sloop of war *Vincennes*, in 1842. Having earned a stellar reputation in the navy, he was given the ship ahead of much more senior, if less competent, officers. The appointment caused considerable controversy, sparking debate over the question of seniority versus merit. It was a fight that would continue for some time, right through the tenure of Stephen Mallory in the Senate and even into the Confederate navy.

For two years "Buck," as he was known to friends (he did not yet warrant the sobriquet "Old Buck"), commanded the *Vincennes* in the Gulf of Mexico. In 1845 he was asked by the secretary of the navy, George Bancroft, to help in founding the United States Naval academy in Annapolis, which he did, becoming its first superintendent. The superintendent's residence on the academy grounds, built in 1905, was in 1976 named the Buchanan House. (The main parade ground nearby is named Worden Field for another former superintendent of the Academy.)

With the outbreak of the Mexican War, Buchanan left the academy for command of the sloop of war *Germantown*. Though the navy's participation in the war was not extensive, Buchanan managed to get himself involved in several sharp fights, leading landing parties against Mexican fortifications.

After the war, Buchanan waited several years for another command, eventually getting the plum assignment of the steam frigate *Susquehanna* in the Far East. He served as flag captain to Commodore Matthew Perry during the opening of Japan, personally conducting much of the negotiations.

Buchanan returned to the United States in 1855 and relinquished command of the *Susquehanna*. He would never again command a ship of the U.S. navy. Soon after his return, he was appointed to Senator Stephen Mallory's controversial navy retirement board.

That same year, Buchanan was promoted to captain and put in command of the Washington Navy Yard, one of the choicest assignments in the navy. Buck was near the pinnacle of his career, one of the most important men in the U.S. navy, with the rank of captain, the highest rank in the service at the time.

Buchanan remained at the Washington Navy Yard through the late 1850s and into the 1860s. He remained there after the Confederates fired on Fort Sumter, even making arrangements for the defense of the yard, including the detonation of the Ordnance Building if that became necessary. "[I]n the event of an attack," he ordered his subordinates, "I shall require all officers & others under my command to defend [the yard] to the last extremity." On April 19, 1861, Buchanan arranged for the *Anacostia* to be loaded with incendiaries for the destruction of the Gosport Naval Shipyard, lest it fall into rebel hands.

Then, on the same day Buchanan was loading the *Anacostia* with combustibles for Norfolk, the 6th Massachusetts Regiment was attacked while marching through Baltimore on their way to Washington. The firefight left four soldiers and twelve Baltimore civilians dead. Buchanan, like most Americans, assumed that Maryland would secede.

Franklin Buchanan was a Marylander and a slaveholder, with no love for abolitionists or Yankees in general. Always a man of action, sometimes rash, Buchanan went personally to Gideon Welles's office on August 22 and tendered his resignation, "the most unpleasant duty" he ever performed.

Welles told Buchanan that he regretted the captain's decision, but he did not try to change Buchanan's mind. In an atmosphere where the secretary of the navy was distrustful even of Southern men who proclaimed their allegiance to the Union, he had no interest in one who was demonstrably pro-South. Buchanan left Welles's office, turned the command of the Washington Navy Yard over to John Dahlgren, and returned to his home and Maryland, where he waited for the state, which he considered "virtually out of the Union" to secede.

But as it happened, Maryland did not secede, and within a few days Buchanan understood that he had acted too hastily. He wrote to Welles attempting to undo his resignation, which he felt was "tendered under peculiar circumstances."

Welles, of course, was not about to reinstate him. In fact, he had never given Buchanan the honor of accepting his resignation. Instead, he sent Buchanan an official notice, saying, "By direction of the president, your name has been stricken from the rolls of the Navy." It was the ultimate and irrevocable censure, wiping out forty-six years of honorable service. Buchanan wrote to a friend, "The deed is done, and I am an unhappy man."

In a letter to fellow naval captain Samuel Du Pont of Philadelphia, Buchanan explained his terrible conflict. He had, he claimed, "always gone strong for the Union, but . . . my feelings are with the South . . ."

The conflict of which Mary Boykin Chestnut spoke, a love of the U.S. navy versus a dislike of Yankees, was strong in Buchanan. "I tell you this *candidly*," he wrote to Dupont, "for my whole *heart* and *soul* were centered in the navy, as you know. I have been as strong a Union man as there is in the State, and opposed to *Secession*, but to follow the fortunes of your state . . . is natural to all men except *Northerners*, they would ascertain first what was to be *gained* or *lost* by holding on . . ." Then apparently recalling to whom he was writing, Buchanan added, "when I speak of Northerners I mean all north of Penn[sylvania]." Buchanan's friendship with Dupont would not last much longer.

Before Lincoln's directive ended any chance of his returning to the U.S. navy, Buchanan had hoped to be reinstated and sent on foreign station. But even that might have ended his career, as Welles considered any officer requesting foreign assignment to be suspect. When it was clear that he would not be returning to the navy at all, Buchanan determined to retire to his Maryland farm and sit the war out, not bearing arms for either side. "I cannot bring myself to join the Southern Navy," he wrote to Dupont in May.

But that reticence was not likely to last long. As his anger at the treatment he received from Welles smoldered, as what he perceived as the outrages of the Lincoln administration escalated and the war grew hotter, Buchanan drifted toward joining with the Confederacy. Still, it was not until September 4, more than a month after First Manassas, that Buchanan presented himself to Stephen Mallory and offered his considerable talents to the Confederate Navy.

Buchanan's resentment at Welles's refusal to reinstate him in the U.S. navy ran deep. More than a year after the battle between *Virginia* and *Monitor*, he wrote to Catesby Jones from Mobile, Alabama, where he was in command of naval defenses, "I am anxious to have another crack at the vile vagabonds. It would please me not a little to sink my old ship, the *Susquehanna*, now off this harbor," a surprisingly bitter note for a sailor, whose tendency generally is to forever love their old ships.

Buchanan's heart, apparently, was no longer divided, and Welles had created in him a dangerous enemy.

Franklin Buchanan was sixty-one years old when he took command of the James River squadron. He was described by a young friend as "far less striking in appearance [than French Forrest], quiet, kindly, and as unpretentious as a country farmer, but with an eye which age had not dimmed, and which even then was filled with the light of battle." He was all but bald on top, with white hair ringing his head and a face that was clean-shaven.

Quiet and kindly might not have been the adjectives Buchanan's subordinates would have used. Old Buck was a hard disciplinarian and did not suffer fools or any breeches of naval protocol. Later in the war he would tell the crew of the former *Arkansas* that he "had no use for them as they had no uniforms!" One of *Virginia*'s officers described Buchanan as "a typical product of the old-time quarterdeck, as indomitably courageous as Nelson, and as arbitrary. I don't think the junior officer or sailor ever lived with nerve sufficient to disobey an order given by the old man in person."

The same officer related a story from the opening of Japan. The *Susquehanna* was going upriver under the command of a Chinese pilot who put the ship aground. Buchanan rounded on the pilot "so fiercely that the Chinaman jumped overboard." But unlike the "Chinaman," the men of the *Virginia* loved Buchanan, and they would stand by him against any odds.

Like John Brooke, Buchanan's work in Richmond was of such critical importance to the navy that it was difficult for him to get away. He wrote to French Forrest, "Mr. M [Mallory] keeps me here to arrange matters, he says I can do more by hurrying people here than by being in Norfolk."

Catesby Jones had born the brunt of the fitting out and training of the crew. Buchanan's appointment made his position more awkward. "Until he was ordered," Jones wrote, "I was listened to, but of course that cannot be now."

Once Buchanan received his orders he traveled to Norfolk to look over his flagship. "I found the ship by no means ready for service," he wrote, "she requires eighteen thousand two hundred pounds of powder for the battery, Howitzers were not fitted and mounted on the upper deck to repel boats and boarder & none of the port shutters fitted on the ship. Much of the powder has now arrived and the other matters shall not detain me."

The history of the Civil War is full of examples of officers who found excuses, valid or otherwise, for not taking action. Buchanan had a perfectly good excuse for not taking action, but he took it anyway. Mallory had found just the right man to command the Confederate navy's ultimate weapon.

Chapter 24

John L. Worden

New York, January 16, 1862.
SIR: I have the honor to report that I have this day reported for duty for the command of the U. S. ironclad steamer building by Captain Ericsson.

Respectfully, your obedient servant,
—*Lt. John L. Worden to Gideon Welles*

In mid-January, *Monitor* got her captain. He was not the first choice. On the 15th, Commodore Joseph Smith wrote to Stimers, "Lt. Jeffers is on duty and cannot now be released. I have applied for Lt. John Worden to command Ericsson's battery and he has been ordered."

Jeffers would get his chance. In fact, he would succeed Worden as *Monitor's* regular captain, a change that would be much lamented by the crew. But at the time of *Monitor's* launch, Jeffers was commanding a sidewheeler with the distinctly unmilitary name of USS *Underwriter*. Lieutenant Jeffers would be *Monitor's* third commander, not her first.

Her first commander was Lieutenant John Lorimer Worden. On the 11th, Smith had written to Worden, "I have only time to say I have named you for the command of the battery under contract with Captain

Ericsson, now nearly ready at New York. This vessel is an experiment. I believe you are the right sort of officer to put in command of her."

Lieutenant John Worden was forty-three years old with twenty-six years in the navy. He was described by one of *Monitor*'s officers as "tall, thin & quite effeminate looking . . . He is white & delicate . . . & never was a lady the possessor of a smaller or more delicate hand . . ." Even in an age when facial hair was the rule, Worden's beard was impressive, coming straight down to midchest.

John Worden would end the war as a hero, and end his navy career as an admiral, but his Civil War started out very badly indeed. He had the distinction—and the bad luck—of being one of the very first Union prisoners of war.

Fort Pickens

In the early months of 1861, while the North and South were sweating out the stalemate at Fort Sumter, they were also wrangling over the similar if less dramatic situation taking place at Fort Pickens in Pensacola, Florida.

Unlike the situation in Charleston, however, the U.S. navy had a presence in Pensacola. A squadron of ships, led by the screw steam frigate *Brooklyn* and the sailing frigate *Sabine,* lay just offshore. Onboard the vessels was an artillery company under the command of Captain Israel Vodges, along with the marine companies from the various ships, ready to land on Santa Rosa Island and reinforce Fort Pickens.

The troops, however, remained on the ships, kept there by the truce put in place during the last days of the Buchanan administration by Welles's predecessor, Isaac Toucey, and the soon-to-be secretary of the Confederate navy, Stephen Mallory. General Braxton Bragg, commanding the Confederate forces in Pensacola, agreed to leave Fort Pickens alone if Captain Henry Adams, commanding the Union naval forces, did not try to land reinforcements. For two months, while the entire country was careening toward war, both sides watched each other but did nothing.

The Lincoln administration was in the same quandary over Pickens as they were over Sumter. In fact, the debates within the cabinet about how to handle the Sumter standoff generally included Pickens as well. Finally, in mid-March, Lincoln decided that Fort Pickens should be reinforced. General-in-Chief Winfield Scott relayed orders to Captain Vodges, who was aboard the *Brooklyn*, instructing him to land his troops at the first favorable moment, then reinforce and hold Fort Pickens.

Vodges wrote to Captain Adams onboard the *Sabine* requesting boats and assistance in landing the troops.

Adams refused. He was still adhering to the truce established by the Navy Department, and was not impressed with orders from the War Department. Just two days before, he and Bragg had conferred and reaffirmed their agreement. Bragg, Adams knew, would consider the landing of troops to be an act of war. What was more, Adams understood the wider implications of such a move on the delicate political balance in the country. It was an applecart he did not care to upset.

On the same day he refused Vodges's request for cooperation, Adams wrote to Welles explaining his position. "While I can not take on myself under such insufficient authority as General Scott's order the fearful responsibility of an act which seems to render civil war inevitable, I am ready at all times to carry out whatever orders I may receive from the honorable Secretary of the Navy."

The dispatch itself would have caused a crisis if it had fallen into secessionist hands. To carry it to Welles, Adams chose one of *Sabine*'s lieutenants, Washington Gwathmey. Adams had used Gwathmey as a courier before, but he was apparently unaware that the lieutenant was a Richmond native, and a secessionist at heart. Gwathmey headed for Washington with the secret communications hidden in a belt he had fastened on under his shirt.

Gwathmey's behavior was an extraordinary example of fidelity to duty, a fidelity that was displayed by many Southern naval officers prior to resigning. He traveled night and day, passing through his native Richmond without stopping, until he reached Washington on April 6. He went directly to Gideon Welles's office and delivered the sensitive letter.

With that duty fulfilled, Gwathmey resigned his commission and accepted dismissal from the U.S. navy. He went on to have an active career as an officer in the Confederate navy. But as long as he was still an officer of the United States, he faithfully executed his orders and did not betray his trust, not even to the enemy he would soon join.

Welles read Adams's letter and was not happy to find that Fort Pickens was still not reinforced. He brought the letter to Lincoln, and both men agreed that orders had to be sent back immediately for Adams to assist Vodges in landing the troops. War was looking inevitable, and they did not want to risk losing Fort Pickens.

The problem was, who to send back with the orders? In those early days, with resignations pouring in and many Southerners still sitting on the fence, Welles was having a hard time gauging men's loyalty. Nor was it just a matter of loyalty—the mission was potentially quite dangerous. Welles asked Paymaster Henry Etting, whose loyalty he did not doubt, if he would go. Etting was not well, but he agreed to undertake the mission if they could not find another officer instead, one whose loyalty was also above question. As it happened, such a man had just arrived in Washington: Lieutenant John Worden.

It was already late in the evening, but Welles did not want to waste a minute. He asked Worden to report to him, and outlined the mission he needed performed. Welles explained the dangers, and told Worden he would have to leave in two hours. Worden agreed immediately.

Welles then wrote orders for Captain Adams:

SIR: *Your dispatch of April 1 is received. The Department regrets that you did not comply with the request of Captain Vodges to carry into effect the orders of General Scott . . . You will immediately on the first favorable opportunity after receipt of this order afford every facility to Captain Vodges by boats and other means to enable him to land the troops under his command, it being the wish and intention of the Navy Department to cooperate with the War Department in that object.*

Welles did not seal the orders. Rather, he handed them to Worden opened and suggested the lieutenant commit them to memory so that he could later destroy them if he felt it was necessary. When he met with Adams, he could explain the situation and deliver the orders verbally.

Early on the morning of April 7, Worden boarded a train heading south. He told no one of his mission, not even his wife, Olivia. It took the lieutenant four full days of travel to reach his destination. As Welles had warned him, he attracted a lot of attention. In that weird twilight between peace and war it was still possible for a U.S. naval officer in uniform to travel through the South unmolested, but just barely. Had he been caught out of uniform, he could have been taken for a spy. He was questioned several times but managed to talk his way out of trouble. Around Atlanta, Georgia, he destroyed the written orders.

Worden arrived in Pensacola around midnight on the 10th. The next day he reported to Confederate general Braxton Bragg for permission to go out to the Union fleet off the bar. He informed Bragg that he was a U.S. naval officer with a verbal dispatch for Adams. Bragg agreed to allow Worden to go, and wrote him a pass. Worden asked Bragg if he would also be allowed to return, and Bragg assured him he would, if neither he nor Adams did anything to violate the standing agreement.

Worden was taken to the small screw steamer USS *Wyandotte*, which was operating closer inshore, but the weather was too rough for *Wyandotte* to pass over the bar and deliver him to *Sabine*. Worden spent the day onboard *Wyandotte* and the next morning visited Fort Pickens. When the seas had at last subsided, Worden was brought out to *Sabine*. It was 11:00 A.M., April 12, 1861, and unbeknownst to anyone there, the Civil War had just started.

Worden met with Adams and delivered the message from Welles. Finally, presented with proper orders from the proper authorities, Captain Adams was ready to act, and he did not hesitate. That very night, after dark, boats from the squadron, loaded with marines and Vodges's artillery company, pulled for Santa Rosa Island. By dawn the next day, Fort Pickens was reinforced, and there was nothing the Confederates could do to stop it. Pickens would remain in Union hands for the entire war.

John Worden was not there to see the orders carried out. Once he had delivered his dispatch, he returned to Pensacola onboard *Wyandotte*. Adams assured him that he did not need to see Bragg again, so he did not, but rather boarded the first train north.

The reinforcement of Pickens was not a complete surprise. The Confederates realized that something was brewing. A week before,

Bragg had asked if he should attack the fort if it was reinforced, and telegrams had been arriving warning him of troops heading to Pensacola. Nor was Worden's mission unknown in the Confederate capitol of Montgomery. On the day that Worden arrived aboard the *Sabine*, Confederate Secretary of War Walker sent a terse telegram to Bragg: "Lieutenant Worden, of U. S. Navy, has gone to Pensacola with dispatches. Intercept them."

By the time Bragg received the telegram, Worden was gone. Bragg replied by telegram:

> *Mr. Worden had communicated with fleet before your dispatch received. Alarm guns have just fired at Fort Pickens. I fear the news is received and it will be reenforced before morning. It can not be prevented. Mr. Worden got off in cars before I knew of his landing. Major Chambers is in the cars. He will watch Mr. Worden's movements. If you deem it advisable, Mr. Worden can be stopped in Montgomery.*

The Confederate government did deem it advisable. Worden was arrested near Montgomery and put in prison.

A Prisoner of Nearly War

After locking him up, the Confederate government tried to find out what Worden had done to deserve it. On the day Worden was taken prisoner, Walker sent a telegram to Bragg, saying, "When you arrested Lieutenant Worden what instructions, if any, did he show you? Did he communicate to you that he had verbal instructions; and if so, what were they? He is here under arrest, and it is important for you to reply fully."

Bragg replied that he had just received a letter from Adams informing him that Fort Pickens had been reinforced by order of the U.S. government. Bragg claimed, "Lieutenant Worden must have given these orders in violation of his word. Captain Adams executed them in violation of our agreement."

But Bragg was not entirely certain. That same day he sent another telegram telling Walker that the reinforcement had been proceeded by signal guns, and that Worden was already ashore by the time the action commenced. "Worden's message may have had no connection with the move."

Certain or not, Worden was a prisoner of war and would remain so. Bragg's implications that Worden had violated his word of honor in lying about the nature of his message made things go harder for him. Worden wrote that Bragg's accusations created "a very general hostile and harsh feeling from the officials & citizens" of Montgomery. He was denied many of the courtesies he felt he was due as a prisoner of war.

Part of the problem, of course, was the very uncertain nature of things. Worden was arrested on the same day Robert Anderson surrendered Fort Sumter. It would be another two days before Lincoln issued his call for volunteers. Worden was a prisoner of war before there was a war. There would not be an official agreement between North and South on the exchange of prisoners for more than a year.

Worden's friends and family were frantic at the news. Reflecting the uncertainty of the situation, Olivia Worden wrote to a friend, "I am extremely anxious to know what course Gov[ernment] will pursue about exchanges of prisoners—they cannot—they will not—I am sure sacrifice my husband by an unwillingness to treat by exchange—I cannot believe that."

Fortunately, the government was willing to treat by exchange, and Worden was very well connected. His wife and brother began a letter-writing campaign, soliciting anyone who might do some good. Among them was Commodore Joseph Smith, who would later appoint Worden to command *Monitor*, and whom Olivia described as "a good friend of Worden's."

Nonetheless, it was seven months before any real progress was made in securing his release. In September, another friend of Worden's, the hulking, bearded Flag Officer Louis Goldsborough, had been promoted to command of the North Atlantic Blockading Squadron, based in Hampton Roads. Goldsborough took up Worden's cause, writing to Brigadier General Benjamin Huger, who commanded Confederate forces in Norfolk. He explained, "Lieutenant Worden sailed with me some years ago, and I am on terms of intimacy with his family. Hence the reason for my deep feeling of interest."

Goldsborough pointed out that there was a Confederate prisoner of war, a Lieutenant Sharpe, a prisoner aboard the USS *Congress** and anxious for exchange. Would Huger consider swapping Sharpe for Worden?

Huger did not have the authority to make such a decision. He referred the matter to Judah Benjamin, then acting secretary of war. Benjamin approved the exchange. Worden was released and allowed to travel to Norfolk.

On November 20, more than seven months after being captured in Montgomery, John Worden stepped onboard the flagship *Minnesota* in Hampton Roads. He carried with him a letter from Huger to Goldsborough asking for Sharpe's release. Goldsborough turned Sharpe over to the Confederates that day, releasing both men from their parole.

John Worden was a free man, and soon the navy would have reason to be thankful for having him back.

---

*Sharpe used his time as a prisoner of war wisely. Soon after his release he wrote a long and detailed intelligence report to the commander of the James River Squadron about the condition and preparedness of the Union ships, based on his observations.

Chapter 25

A Novelty in
Naval Construction

When Mallory ordered Franklin Buchanan to command the
James River Squadron, he gave him no specific instructions
as to how he was to use the ironclad and her consorts.
What he gave instead was a lengthy analysis of what he believed *Virginia* could do, and what he hoped to get from her.

> The Virginia is a novelty in naval construction, is untried, and her
> powers unknown, and the Department will not give specific orders as
> to her attack upon the enemy. Her powers as a ram are regarded as
> very formidable, and it is hoped that you may be able to test them.
>
> Like the bayonet charge of infantry, this mode of attack, while
> the most distinctive, will commend itself to you in the present
> scarcity of ammunition. It is one also that may be rendered destructive at night against the enemy at anchor.
>
> Even without guns the ship would be formidable as a ram.
>
> Could you pass Old Point and make a dashing cruise on the Potomac as far as Washington, its effect upon the public mind would
> be important to the cause.
>
> The condition of our country, and the painful reverses we have

just suffered, demand our utmost exertions, and convinced as I am
that the opportunity and the means of striking a decided blow for
our Navy are now for the first time presented, I congratulate you
upon it, and know that your judgment and gallantry will meet all
just expectations.

   Action—prompt and successful action—now would be of serious
importance to our cause . . .

The secretary was putting a great deal of his faith in the ironclad's ability, perhaps too much. The letter illustrates how large the ship loomed in Mallory's strategic thinking.

   Mallory was certainly correct in his belief that an attack on Washington would do much for the Confederate cause, perhaps even more than he imagined. The attack on Hampton Roads caused panic enough. He was also right about her powers as a ram.

   He failed, however, to understand the ironclad's limitations—her deep draft, unreliable engines, and prodigious consumption of coal, any one of which would have prevented her from going up the Potomac. Right up to the end he would continue to believe she was capable of more than she was.

   French Forrest had been scrounging needed supplies for the *Virginia* for several months. When he heard that a Union ship had gone ashore near Norfolk with several hundred gallons of oil on board, he wrote to General Hugar requesting that the oil be sent to the shipyard. "[We] are without oil for the *Merrimack*," he wrote, "and the importance of supplying this deficiency is too obvious for me to urge anything more in its support."

   As she lay alongside the dock, *Virginia* received her finishing touches. Mrs. Eliza Young and four other Norfolk women were contracted for curtains and "11 mattresses @ .60¢ each" and "6 boat cushions for CS Steamer *Virginia*." But Forrest's chief concern was gunpowder, the 18,200 pounds of it that Buchanan needed. Letters flew around the Norfolk and Yorktown area as Forrest requested powder from the army. "The *Virginia* is now detained for powder," he wrote to Hugar. ". . . I write to suggest that if you feel authorized to make the transfer from Forts Norfolk and Nelson of the necessary ammunition, it would relieve us greatly." Forrest would repay the army with the powder meant for *Virginia* when it finally arrived.

On February 28 he wrote to Colonel S. Anderson at Norfolk requesting powder from him, as Richmond had sent only 1,000 pounds of the 18,000 needed. Forrest encouraged Anderson to hurry, since it would take three days to fill the ship's cartridge bags. There was a sense of urgency now. He ordered the commander of the receiving ship *States* to save powder for fifty rounds per gun for his ship and send the rest to *Virginia*.

Buchanan, too, was thinking about his flagship. Like Mallory, he saw great potential. His vision involved a joint land/sea operation with General James Bankhead Magruder, commanding the Confederate troops in Yorktown, about 15 miles north of Newport News. During the last weeks of February and the beginning of March, Buchanan kept up a correspondence with Magruder concerning the possibilities of a cooperative effort.

Finally, on March 2, Buchanan wrote to the general, "It is my intention to be off Newport News early on Friday morning next . . . My plan is to destroy the frigates first, if possible, and then turn my attention to the battery onshore. I sincerely hope that acting together we may be successful in destroying many of the enemy."

On the same day that Buchanan was outlining his final plans to Magruder, Magruder decided that he did not want to participate. Mallory wrote to Buchanan, "I am requested by the Adjutant-General to inform you that it will be impossible for General Magruder to act in concert with or render you any aid in the plans agreed upon to attack the enemy at Newport News."

Magruder had plenty of excuses, primarily that the roads were too muddy to move his troops. But the real reason for his hesitancy was that he did not think the plan would work. "I am . . . satisfied that no one ship can produce such an impression upon the troops at Newport News as to cause them to evacuate the fort," he informed the War Department.

Virginia did, however, mete out real punishment to the Yankee fortifications. Her broadsides were "very destructive to the shore batteries." Some batteries were abandoned, and many federals assumed Magruder would be attacking, and that he would succeed in taking Newport News. But in the end, Magruder made no serious effort against the Federal positions there.

With Magruder out of the picture in terms of planning the action, Mallory instructed Buchanan to act independently, which he had every intention of doing. He officially came aboard *Virginia* on Tuesday, March 4, and hoisted his red admiral's pennant on the forward flagstaff. It had been his intention to attack the Union squadron the following Friday morning, and Magruder's refusal to cooperate would not change that. "On Thursday night the 6th inst. I contemplate leaving here to appear before the enemy's ships at Newport News," he informed Mallory. To quash the secretary's excessive enthusiasm, he added, "From the best and most reliable information I can attain from experienced pilots it will be impossible to ascend the Potomac."

Virginia was about as ready as she was going to be. Buchanan expected to get the last of the powder and shell onboard the next day, Wednesday, and the howitzers on Thursday, though there is no indication that the howitzers were ever mounted.

It was originally planned that each of *Virginia*'s gun ports would be fitted with a 3- or 4-inch-thick hammered iron shutter to shield the men inside from enemy shot. The shutters were oval-shaped and slit down the middle, so they would open and close like a pair of scissors, operated by chains inside the casemate. A round opening in the middle of each shutter would allow them to close around the barrel of a gun run out through the port. But with the Tredegar's bottle-neck of iron production, the shutters were put on the back burner. They were not in place when Buchanan came aboard.

The admiral wrote to Mallory, "The shutters for the two bow & quarter ports I will have temporarily placed." The bow and stern pivot guns, on Brooke's suggestion, had three ports rather than one. That meant there would, at any given time, be two ports gaping open, without even a cannon barrel to interfere with enemy shot coming into the casemate. Accounts differ, but apparently only the shutters on the two bow ports and two quarter ports were in place when *Virginia* sailed.

A Bad Piece of Work

Despite Forrest's additional security at the yard, word of *Virginia*'s problems leaked out and was published in an article in the *Norfolk Day Book*. The story outlined the ship's problems with her displacement, claimed her plating was insufficient and her machinery all but worthless. The paper assured readers that she would at least be invaluable as a floating battery. The *Mobile Register* picked up the story, calling the ship a "bad piece of work."

About a week before *Virginia* left Norfolk for the first time, an escaped slave, described as an "intelligent contraband," slipped out of Norfolk and made it out to the ships blockading the James River. He brought with him a copy of the *Day Book*. Apparently the Union navy men (and some Confederates) thought the article was a disinformation campaign, intended to get them to let their guard down. They dismissed the claims.

The article certainly made *Virginia* out to be worse than she was, but whether it was intentionally hyperbolic or not is hard to say. *Virginia was* a deeply flawed ship. Catesby Jones, the officer who knew her best, said she was "unseaworthy, her engines were unreliable, her draft, over twenty-two feet, prevented her from going to Washington."

Worst of all, the ship still rode too high in the water. Jones had been ordered to put no more ballast in her, for fear of straining the bottom, but she was still a foot higher than she was supposed to be.

"The eaves of the roof will not be more than six inches immersed," Jones complained to Brooke just a few days before the first day's fight, "which in smooth water would leave it bare except the one-inch iron that extends some feet below." One inch of iron offered little protection, a particular problem at the waterline, where a well-placed ball could sink the ship. Jones speculated that a "thirty-two pounder would do it."

These details would not stop Buchanan, but the weather would. On Thursday, a huge spring storm swept in, blanketing the East Coast, a "hard gale of wind" that obliged the ships in Hampton Roads to double up their anchors. The unseaworthy ironclad remained tied to the dock, and Buchanan endured the frustration. With the unexpected delay, yard workers swarmed back over the vessel, taking advantage of every moment to get the ship closer to completion.

The next day the storm abated and Buchanan was ready to go. The sides of the ironclad were slushed with tallow—beef fat—which, like tar, had many uses on shipboard. In this instance it was used to "increase the tendencies of projectiles to glance," which may have been the first, and possibly only, time it was put to that use.

Buchanan wanted to surprise the Federals. He wanted to be in Hampton Roads when the sun came up, but that meant steaming down the Elizabeth River at night. The usual aides to navigation, such as buoys and lights, had been removed to make it more difficult for the enemy to get up to Norfolk. To aid the pilots, lights were placed on the various obstacles that littered the river. *Virginia* was once again ready to go.

But when night came, the pilots, five in all, declared that they could not navigate the ship downriver in the dark. In their defense, Jones correctly pointed out that "it was not easy to pilot a vessel of out great draft under favorable circumstances." *Virginia*'s pilots often displayed what many considered an excess of caution. It was this caution that would doom the ship in the end.

The storm passed on Friday night and Buchanan wished to get under way at dawn the next day, but the pilots would not have that either. The tide was out at that hour, and the deep-draft *Virginia* could not make it up the river at low water. They would have to wait for the flood to begin.

The James River Squadron

Virginia, of course, was not the only ship waiting. While her attack on the Union ships, and her subsequent battle with *Monitor*, have been looked on as single-ship actions, since the ironclads so dominated the fighting, in fact both were fleet actions. "The Confederate accounts of the battle were full of the *Merrimac*," complained the commander of one of the gunboats, "and but little was said of the smaller vessels whose fire was equally effective." The Battle of Hampton Roads would pit the Confederates' James River Squadron against elements of the Union's North Atlantic Blockading Squadron.

Buchanan's squadron included the flagship *Virginia* and five other vessels, the *Patrick Henry, Jamestown, Teaser, Raleigh,* and *Beaufort.*

Patrick Henry, commanded by the very able John Tucker (the former ordnance officer at the Gosport Naval Shipyard), was a schooner-rigged sidewheel steamer of 1,300 tons. Formerly the *Yorktown,* she was seized by the state of Virginia following secession and sold to the Confederate navy. *Patrick Henry* mounted ten guns, making her by far the most significant vessel after *Virginia.* The following summer she was designated the school ship for the Confederate navy's naval school in Richmond.

The *Jamestown* was also a sidewheeler, mounting two guns and commanded by Lieutenant Commanding J. S. Barney. *Teaser,* Lieutenant Commanding W. A. Webb, and *Beaufort,* Lieutenant Commanding William Harwar Parker, were both former tugs, the later an iron hull. Each mounted a 32-pounder on the bow. *Raleigh,* Lieutenant Commanding J. W. Alexander, was described as a steam gunboat, carrying one gun.

Of these ships, only *Raleigh* and *Beaufort* were in Norfolk. The other three were in the James River, under the command of John Tucker. Buchanan had written to Tucker earlier, outlining his plans and instructing Tucker to prepare to join the others. "You will, in the absence of signals, use your best exertions to injure or destroy the enemy, much is expected of this ship and those who cooperate with her."

Those little gunboats would be playing with the big boys, steaming into the kind of gunfire few naval vessels could stand up to. But in the bold David-and-Goliath tradition of the Confederate navy, they did not flinch.

A Trial Trip

At around 10:00 A.M. on Saturday, March 8, Franklin Buchanan came on board his flagship. *Virginia* was still swarming with workmen trying to complete the ship, but Old Buck ordered them all off.

Still onboard were the 320 or so crew members: navy men, soldiers, yard workers recruited from Gosport, even a detachment of the United

Artillery of Norfolk, serving as one of the gun crews. "The officers and crew were strangers to the ship and to each other. Not a gun had been fired, hardly a revolution of the engines had been made," one officer recalled. It was far from an ideal situation, but, like the unfinished ship herself, it would have to do.

Buchanan ordered the ship under way, but few men aboard knew where they were bound. Most of the crew believed they were going out for a trial trip, a sea trial, a reasonable assumption considering that the ship had never even left the dock. Some of the officers knew differently.

H. Ashton Ramsay was summoned from the engine room where he was getting up steam in the boilers. He found Buchanan on the shield deck, the roof of the casemate. The captain was "pacing the deck with a stride I found it difficult to match, although he was then over sixty and I but twenty-four."

Buchanan wanted to know about the engines, the most worrisome aspect of his new command. "Ramsay, what would happen to your engines and boilers if there should be a collision?" Buchanan was already thinking about the "formidable" properties of the ram.

"They are braced tight," Ramsay assured him. He explained to Buchanan how the joints in the steam pipes were made flexible to withstand shock.

"I am going to ram the *Cumberland*," Buchanan revealed. "I'm told she has the new rifles guns, the only ones in their whole fleet we have cause to fear. The moment we are in the Roads I'm going to make right for her and ram her. How about your engines? They were in bad shape in the old ship, I understand. Can we rely on them? Should they be tested by a trial trip?"

Ramsay knew all too well how bad the engines had been, but he had no way of knowing how much good he and Williamson had done, since *Virginia* had never been under way under her own power. But Old Buck didn't care for equivocation and Ramsay knew it. "She will have to travel some ten miles down the river before we get to the Roads," Ramsay said optimistically. "If any trouble develops I'll report it. I think that will be sufficient trial trip."

By 11:00 A.M. the tide was well into the flood, steam was up, and the men were ready. Fore and aft the dock lines were cast off, and the mighty ironclad *Virginia* moved out into the stream. The last of the navy yard's workmen leapt ashore as she pulled away from the dock. Hovering nearby, like pilot fish to a shark, were the former tugs, *Raleigh* and *Beaufort*. The sun was shining, the water calm, the day perfect.

Fifteen miles away, the Union ships *Congress* and *Cumberland* and their hundreds of sailors enjoyed the weather and their easy duty, anchored at the mouth of the James River. While the *Virginia* was getting up steam and preparing to get under way, the sailors were washing their clothes and running them up in the rigging to dry.

Within five hours, hundreds of them would be dead, the U.S. navy would suffer the greatest defeat it had ever known, and the whole world's idea of what a man-of-war was would forever be altered.

The *Virginia* was coming out.

Chapter 26

Testing Her Capabilities

Worden's time in a Southern prison left him weak and unwell, with a fever that would recur. His physicians and family urged him not to take command of the *Monitor*. But he did, and greeted his new assignment with enthusiasm.

When Smith first intimated that he might have command of the Ericsson Battery, Worden inspected the vessel while she was still on the ways. He wrote enthusiastically to Smith, "After a hasty examination of her, am induced to believe that she may prove a success. At all events, I am quite willing to be an agent in testing her capabilities, and will readily devote whatever of capacity and energy I have to that object."

On the same day that Worden made his inspection, Welles officially appointed him the *Monitor's* "Lieutenant Commanding." Worden assumed command on January 16, 1862, fourteen days before the vessel's launch.

With the ship still being built, Worden delved into the other concerns of his command, in particular filling out the complement of officers and crew. He consulted with Stimers regarding how many men would be needed to fight the ship. Worden explained to Welles, "In estimating the number of her crew, I allowed 15 men and a quarter gunner for the two guns, 11 men for the powder division, and 1 for the wheel, which I deem ample for the efficient working of her guns in action. That would leave 12 men (including those available in the engineer's department) to supply deficiencies at the guns, caused by sickness or casualties."

Worden was concerned about the close quarters and the ventilation of the turret. Seventeen men and two officers was his estimate for the number of men who could work in that space. Any more, he felt, "would be in each other's way and cause embarrassment."*

A few of the ship's officers had arrived before Worden. First Assistant Engineer Isaac Newton reported for duty on December 7 to fill the role of chief engineer. Newton had worked under Alban Stimers onboard the USS *Roanoke*, and Stimers was well pleased with him. Like his mentor, Newton was a great ironclad enthusiast, and felt that the Ericsson Battery was the apex of ironclad development. He had nothing but disdain for ships such as *Warrior* and *Le Glorie*, sailing vessels with auxiliary steam and iron bolted to their sides. He considered it an advantage that the United States was late getting into the ironclad game, and could learn from the bad example set by the British and French.

With more insight into the future of naval warfare than many of his older colleagues possessed, Newton wrote, "In all vessels now being built for the navy, speed under steam is *sine quo non*; the hallucination of *auxiliary steam power* has been exploded." An assignment to *Monitor* under Alban Stimers must have been a dream job for Newton.

Another officer to proceed Worden was Acting Assistant Paymaster and Clerk William Keeler, a newly minted officer in the Volunteer Navy, which had been created as a reserve force for the standing navy, much as had been done in the army. Keeler was from Illinois, and his seagoing experience was limited to a voyage as a passenger around Cape Horn to San Francisco, where he had participated, without success, in the Gold Rush. Keeler was also senior partner at the firm of La Salle Ironworks, Founders & Machinists. His familiarity with machinery would put him in good stead when things started to go wrong in the *Monitor*'s engine room.

With his round wire-frame glasses, mustache, and long chin beard,

---

*An examination of the recently excavated turret makes one wonder how even nineteen men could work in that space without falling all over one another.

Keeler looked the part of a clerk, a job he performed well. Of greater significance, at least for the historian, Keeler faithfully wrote letters to his wife, Anna, which she faithfully preserved. Keeler meticulously described life aboard the ironclad, from the high points of battle and savage weather to the mundane of everyday existence. For much of *Monitor*'s human history, we are indebted to Paymaster Keeler.

Monitor's executive officer, the second-in-command, was Samuel Dana Greene. Greene was only twenty-two years old, a handsome man with dark hair, a thick mustache, and piercing eyes. He graduated from the naval academy in 1859 and served as a midshipman in the China Squadron onboard the *Hartford*, the ship that Farragut would make famous during the Civil War. Even before he was assigned to *Monitor*, the ironclad fascinated him, and he took every opportunity to examine her during her construction. By the time he volunteered for *Monitor* duty, he was thoroughly familiar with her design.

Greene served on the *Monitor* at Worden's request, as did many of the officers. On January 23, Commodore Smith wrote to Stimers informing him of Greene's appointment. "Lt. Green [sic] will be ordered today. Lt. Worden applied for Lt. Green." Then, in Smith-like fashion, even at that late date, he could not help but add, "A good deal of wonderment and many surmises of failure are afloat."

Greene arrived in time to join Worden and Ericsson on the *Monitor*'s deck as the vessel slid down the ways. Though *Monitor* would have five commanders in her short life, Greene would be the only man to ever serve as executive officer. He was onboard her when she first touched water, and he was among the last to leave when she slid beneath the waves off Cape Hatteras less than a year later.

For enlisted men, Worden was given leave to select sailors and black gang from any of the men-of-war in New York Harbor. He went first to the *North Carolina*, a sailing ship of the line launched in 1820, a throwback to the earlier era of naval warfare that was being ushered out by the *Monitor* and the *Virginia*. Decrepit and obsolete, the *North Carolina* was being used as a receiving ship, a place to house recruits waiting for assignments, and a none-too-pleasant one. One resident described her as "a devilish old hulk."

Worden also went onboard the frigate *Sabine*, which he had visited on his courier mission to Pensacola. On each ship he explained the situation. Worden clearly had doubts about *Monitor*'s seagoing capabilities. He told the men, "I won't draft any of you for service on that thing, I merely call for volunteers. I can't promise to get you to Hampton Roads, but if I ever do I think she will do good service."

Despite the less than inspiring call to arms, far more men stepped forward than the forty-two Worden needed, and he was able to pick the cream of both ships. He would later say of the crew, "A better one no naval commander ever had the honor to command."

Ironclad Ordnance

After *Monitor*'s launch, John Ericsson's frenetic efforts dropped off dramatically. His part was largely done; the rest was up to the navy. And the navy was pushing hard to get her to sea, with the crew working long hours on the myriad unfinished details.

One of the ongoing issues was the question of the guns that would be mounted in the turret. Though Ericsson's original specifications called for 12-inch guns, it was really his wish to see 15-inch guns mounted in his battery. But guns of that size were not available. In fact, the only guns the Ordnance Department could supply were 11-inch, and those would not be available until the vessel was near completion.

Stirring things up even further, Ericsson suggested that the guns be shortened by 18 inches so they could be more easily loaded in the confines of the turret. Stimers took up the cause. Since the ironclad was impregnable, he reasoned, her commander would always be fighting at the closest range possible. Therefore, "the greater accuracy at long ranges, secured by the greater length, would very seldom be available, whereas, the convenience and consequent rapidity of loading the shorter length, would add to the effectiveness of the vessel at every discharge of the guns."

Commodore Smith did not buy it. He worried that Ericsson had miscalculated, that the turret was in reality too narrow to work the full-length guns. Ericsson assured him that was not the case. Finally Smith wrote, "My object and my pride is fixed on those XI inch guns. I cannot agree to shorten these and would not like to ask Dahlgren's consent to do so, knowing he would object to it."

Commander John Dahlgren, the Union's ordnance genius and the inventor of the 11-inch guns Monitor would carry, had many concerns about the weapons. He ordered that in no case should shot weighing more than 170 pounds be fired from them. And, worried about the possibility of their bursting, he ordered that they not be loaded with more than 15 pounds of powder, half the charge that an 11-inch Dahlgren was designed to take. Many people, including Worden, would claim, with justification, that it was precisely that order that prevented Monitor from tearing Virginia's casemate apart.

At the time of the Civil War, one of the most significant developments in the area of naval ordnance was the exploding shell fired from a rifled gun. Accurate at ranges that smoothbores could never attain, with explosive power devastating to wooden vessels, these new weapons were ushering in the age of naval combat at long range, and spelling the end of the old style of fighting "yardarm to yardarm." But the Monitor mounted smoothbores, and she went into her fight with solid wrought-iron shot ordered by Stimers specifically for that ship.

The reason had to do with Monitor's mission. In explaining the decision to use smoothbores and round shot, Isaac Newton wrote, "Spherical shot are much more effective at short ranges than rifled shot at any range, and as this vessel is shot-proof, she will engage the enemy at a distance of from 300 to 400 yards." Newton's estimate was ten times farther that the actual distance at which Monitor engaged Virginia, but his reasoning, the same that Stimers used to advocate shortening the guns, was correct.

Exploding shells could penetrate the sides of a wooden ship and detonate with terrible consequences. That was why Virginia, going after the wooden fleet, was provided almost exclusively with such ordnance. But the same shells would break up into useless fragments when fired at an ironclad, and Monitor was built to fight an ironclad. What was needed was the punch of solid shot, shot made of wrought iron, not cast iron, which tests had shown would shatter against an impregnable target. And the shot needed the force of 30 pounds of powder behind it.

Trials and Tribulations

On February 17, steam was applied to the auxiliary engine driving the turret, and the turret's movement tested. It turned just as Ericsson designed it, whipping around at two and a half revolutions per minute, and easily controlled by a single operator. The ship was now more than a month overdue, but every indication was that she would soon be ready for sea.

On February 20, Welles sent orders to Worden to "[p]roceed with the USS *Monitor,* under your command, to Hampton Roads, Virginia." The U.S. navy was living on borrowed time with regard to the CSS *Virginia,* and they were growing increasingly desperate for *Monitor* to get under way.

The order was a bit premature—*Monitor* had not even been commissioned yet—but she was the Union's answer to *Virginia* and they needed her on station, because everyone was certain that the Confederate ironclad would be out at any moment. On February 17, David Porter assured Fox that the *Merrimack* would be out within a week. The day after Welles sent the order for *Monitor* to go to Hampton Roads, Captain John Marston, the senior captain at Hampton Roads, informed Welles that "the *Merrimack* will positively attack Newport News within five days." The actual attack was still three weeks away.

The level of concern had been so high for so long that some officers were growing weary of it. The captain of the *Minnesota* wrote, "The *Merrimack* is still invisible to us, but report says she is ready to come out. I sincerely wish she would; I am quite tired of hearing of her."

The day before Welles ordered *Monitor* to Hampton Roads, the ironclad got under way for the first time, steaming down the East River. It was clear from that trial run that she was not going anywhere for a while.

Instead of the eight knots that Ericsson had promised, she was barely able to make three and a half. Stimers discovered that the superintendent at Delameter Iron Works (whom Ericsson had described as "too stupid to make a blunder") had set the engine's cutoff valves for backing instead of going forward. The end result was that the amount of steam being admitted to the cylinders was being cut off at half the stroke of the piston, which greatly reduced the power available for the propeller. To further exacerbate the problem, a valve of one of the blower engines came off its stem, putting the blower out of commission and reducing by half the draft to the boiler fires.

Monitor limped back to the Brooklyn Navy Yard and anchored for the night. The next morning a tug towed her to the wharf. The press, which had been watching the vessel closely, and had been disappointed when she did not sink at launching, went to town on her now. They dubbed her "Ericsson's Folly" and lambasted the government for wasting so much money on so clearly absurd a project.

Stimers called on Ericsson and informed him of the problems. Ericsson was not happy that his vessel proved to be less than perfect. But the imperfections were minor, and had nothing to do with the design, and in a few days they were corrected.

Finally, on February 25, the ironclad was commissioned and officially became the United States Ship *Monitor*. But it was provisional. The Navy Department was still delinquent in their payments, still held their 25 percent, and the vessel still had to undergo her trial by fire before the navy would ultimately agree to accept her.

On commissioning day, the crew moved onboard. Heaters had yet to be installed, and the ship was miserably cold. Fireman George Geer (who, like Keeler, would leave the world with a magnificent series of letters to his wife) wrote, "It is so cold we most freese nights and I am most frose writing this Lettor." After his first night onboard, Keeler "hurried into the engine room to thaw out." By early March, heaters were in place and the vessel was much more livable.

The day after her commissioning, the problems with *Monitor's* engine had been corrected and she was ready to proceed to Hampton Roads in accordance with the increasingly frantic orders coming from Washington. All that was left to do was bring her ammunition onboard, which consumed all of that day. The next morning, February 27, came in with thick weather and a heavy snowstorm. Despite the weather, Worden took on a pilot in the early morning and got under way, anxious to get to Virginia waters. Once again, *Monitor* did not get far.

This time it was her steering. As the helmsman tried to turn the ship, the wheel became unmanageable, slamming hard over and staying there. When the helmsman did manage with great difficulty to turn the wheel, it would slam over to the other side. The *Monitor* careened from bank to bank down the East River, barely under control. As Keeler described it, "We ran first to the New York side then to the Brooklyn & so back & forth across the river, first to one side, then to the other, like a drunken man on a sidewalk, til we brought up against the gas works with a shock that nearly took us from our feet."

It was clear to Worden that the ship could not go to sea. Once again she was towed back to the Brooklyn Navy Yard.

Initially some thought that the wheel was too small to give the helmsman sufficient leverage, but the defect was more involved than that. The root of the problem was a poorly balanced rudder. Too much of the rudder's surface was forward of the rudder post—the rudder's pivot point—in proportion to the area aft. When the rudder was turned from the centerline of the ship, the pressure of the water on the area forward of the rudder post forced the rudder hard over and held it. The force was so great that the man at the wheel was barely able to overcome it. On the *Monitor*'s first trial run, with her speed barely three and a half knots, the problem was not noticed. When her speed, and thus the flow of water around the rudder, was doubled, the steering gear became unmanageable.

The navy men involved with the project felt that a new rudder was the only solution. They insisted that Ericsson haul the ship out and replace the rudder he had originally designed for her.

Ericsson was furious. This was meddling on an intolerable level, particularly considering that the navy had not even had the good grace to pay for the ship. As he wrote to a friend after the fact, "I was all but ordered by the Government officials to take the vessel to the dock to *change the rudder*. Knowing the perfect proportions of it I got so angry at being thus importuned that I told them I would be damned before I did."

Considering the weeks of delay that a new rudder would have meant, it was fortunate that Ericsson dismissed the others and came up with a quicker solution. Wire tiller ropes ran from the steering wheel aft to an arc-shaped tiller attached to the rudder. Ericsson detached the wire tiller ropes from the tiller and in their place put blocks (pulleys), then ran the tiller ropes through the blocks and made them fast to a deck beam. The increase in purchase was enough to overcome the problem with the rudder's balance, and two days later the vessel was ready to go.

Worden was not convinced that Ericsson's solution would do the job. Before he took the *Monitor* to sea, he insisted that she be put through a trial run. He also insisted that a board of three officers come along on the trial to determine if simply adding two blocks to the steering gear was a sufficient fix.

It was March 3 before *Monitor* was ready to try again. With a commission onboard consisting of a commodore, a chief engineer, and a naval constructor, the *Monitor* raised anchor and once again steamed down the East River. Ericsson's fix worked perfectly. The man at the wheel had complete control over the vessel. With the propeller turning at fifty rotations per minute, the helm was put hard to starboard and the ship turned 180 degrees in four minutes and fifteen seconds, and did it in just three times her own length, a turning radius that the *Virginia* could only dream about.

The trial run for the improved steering gear was also an opportunity for the men to fire the 11-inch Dahlgrens (known to the navy as numbers twenty-seven and twenty-eight, West Point, 1859 though called simply one and two on board *Monitor*) for the first time in *Monitor*'s turret. First a blank cartridge was tried in each gun, then a stand of grapeshot and last a round of cannister shot. The gun trial was a success, though it came within an inch of being a disaster.

Alban Stimers was in the turret and working the recoil mechanisms. Ericsson had provided *Monitor's* gun carriages with a friction device to dampen the recoil, such as he had first used on *Princeton*. A handwheel was turned to create more or less friction on the gun's track and thus control the distance the gun would recoil. What Stimers did not know was that the screws to which the handwheels were attached had a lefthand thread, which meant they had to be turned counterclockwise to tighten. Stimers twisted the hand wheel on gun Number One clockwise, confident he was applying more friction, when in fact he was doing just the opposite.

The gun was loaded, the gun captain jerked the lanyard, and the gun rocketed back on its track, stopping only when it slammed into the side of the turret.

Stimers realized his mistake, or at least he thought he did. Unfortunately, since the guns were mirror opposites, he seemed to have thought the friction screws were opposites as well, meaning that Number Two should have a righthand thread, which it did not. Once again Stimers turned the wheel in the wrong direction, and once again the gun flew down the length of the track and smashed its cascabel into the side of the turret. Amazingly, no real damage was done, though one can imagine the shaken nerves in the *Monitor*'s turret.

Monitor returned to the yard under her own power, making six and a half knots against a strong ebb tide. Though most of the crew were pleased with the vessel's performance, at least two men concluded that ironclad sailing was not for them. The log entry for March 3, reads, "Norman McPherson & John Atkins deserted taking the ships cutter & left for parts unknown."

Despite the shakeup in the turret, the test proved satisfactory. The ship was complete. A little over six months from the moment John Ericsson pulled his model from its dusty box and showed it to Cornelius Bushnell, USS *Monitor* was ready to go to war.

Chapter 27

March 8

Franklin Buchanan knew where he was going, and he was going for the sailing men-of-war blockading the James River. It was a little after 2:00 P.M. when the *Virginia* and her consorts, *Beaufort* and *Raleigh,* opened up on the enemy. *Virginia* poured her broadside into *Congress* as she steamed past, doling out horrible carnage, then made for *Cumberland.*

With the momentum of 3,200 tons steaming at six knots, Buchanan drove his ship's cast-iron ram into the frigate's side. For a terrible moment, *Cumberland* and *Virginia* were locked together, the one dragging the other down. *Virginia*'s engines, underpowered and unreliable, labored to pull the ironclad free.

In the engine room, Ramsay had the throttle wide open as he and the black gang listened to the thud of ordnance hitting the shield one deck above. Suddenly there was a tremendous crash in the fire room, as if a boiler had exploded. One of *Cumberland*'s shells had detonated in the smokestack.

Still, *Virginia* would not move. *Cumberland*'s entire weight "hung on our prow and threatened to carry us down with her," Ramsay recalled. Then, finally, with a wrenching sound, the *Virginia* pulled herself free and backed away from the dying ship. Her cast-iron ram had torn clean off and remained in *Cumberland*'s side.*

On *Cumberland*'s deck, Selfridge watched the ironclad pull away and knew that an opportunity had been lost. *Cumberland*'s starboard anchor had been hanging directly over *Virginia*'s foredeck. If it had been let go, it might have held the ironclad where she was, and both ships would have gone down. But that was not done.

Now *Virginia* pulled free and the tide twisted her in the stream. To Selfridge she appeared stunned, lying momentarily motionless and exposed to *Cumberland*'s guns. It is testament to the discipline, training, and courage of the *Cumberland*'s men that even in that circumstance— their ship sinking fast, the decks heaped with dead and wounded, running with blood—the crews still stood by their guns and flung shot at an enemy whose victory was already assured.

Cumberland's men managed to get off three broadsides in quick succession, fired from not more than 100 yards away, and for once the shot was effective. A shell struck on the edge of *Virginia*'s open bow gunport. One of the men on the gun crew who had stuck his head out of the port was instantly killed. Shrapnel tore through the interior of the casemate, killing another man and wounding several others.

The aftermost 9-inch Dahlgren was also struck just as it was being run out. The end of the muzzle was broken off. Another gun had its muzzle shot off as well, leaving the barrel so short that every time the gun was fired, the muzzle flash set the gun port on fire.

The men on *Virginia*'s gun deck did not pause any longer than it took to drag the wounded and dead out of the way. The lieutenants kept the gun crews at their work.

"Sponge! Load! Fire!" the orders were shouted out.

"The muzzle of our gun has been shot away!" one of the gunners called.

"No matter," shouted Jones, who seemed to be everywhere at once, "keep on loading and firing—do the best you can with it!"

---

*In defense of the workmanship that went into *Virginia*, John Luke Porter claimed that the ram had broken in two, but was "so well secured . . . the fastening to the vessel were not broken loose." Most other witnesses say it was pulled right off.

"Keep away from the side ports, don't lean on the shield, look out for sharpshooters!" Men who leaned against the casemate were knocked flat by the concussion of shells on the iron and were carried below, bleeding at the ears.

A number of observers, including Selfridge, who had a pretty good view of things, claimed that *Virginia* rammed *Cumberland* a second time, "for some reason, which I have never understood," he wrote, "unless to give a final *coup de grâce.*" Others, including Buchanan and *Virginia's* other officers, do not mention ramming a second time. Certainly there was no reason to, and *Virginia's* already damaged bow would have suffered from such an attempt. Most likely Buchanan decided correctly that one good ramming was enough.

The damage done to *Virginia* by the dying *Cumberland* was the most serous injury that *Virginia* would suffer. Lieutenant Jones was only happy that *Cumberland's* fire was aimed at the gun ports and not the waterline, an Achilles' heel of which the Yankees were not aware.

Cumberland was doomed, but she went down fighting, and with her colors flying. Even as she was settling by the bow, her crew hoisted the gunpowder out of the forward magazine and moved it aft for use in the after guns. They were helpless to aid the wounded men on the berth deck below, and could only listen as they cried out in horror at the water rising around them.

The deck was shattered, and had the look of a slaughterhouse, "covered with the dead and wounded and slippery with blood." Selfridge's first division was so decimated that he could not muster a single gun crew out of the six he had commanded. Still, the men manned the pumps and fired on the enemy. Selfridge led his men forward, hoping to move the number-one gun to a place where it would bear on the ironclad.

By this time the ship was going down fast and the powder that was accessible was used up. *Cumberland* took a hard roll to port and the water began to pour in through the gunports. It was hopeless. The order was passed, "Every man to look out for himself."

Lieutenant Morris, commanding *Cumberland* in the captain's absence, explained in his report, "Timely notice was given, and all the wounded who could walk were ordered out of the cockpit, but those of the wounded who had been carried into the sick bay and on the berth deck were so mangled that it was impossible to save them."

The survivors clambered down into the boats that had been lowered and tied alongside prior to the fight. Others climbed up in the rigging to escape the rising water, or found gratings or hatch covers or anything that would float. In less than forty-five minutes, *Cumberland* had gone from a powerful man-of-war to a sinking wreck.

The bravery and professionalism with which *Cumberland*'s men fought did not go unnoticed by the Confederates. "No ship was ever fought more gallantly," Lieutenant Wood claimed, and he was by no means alone in that sentiment. One of *Virginia*'s sailors wrote, "The U.S. government should never be ashamed of the action of her officers and men on that memorable day they fought nobly."

Fleet Action

While *Virginia*'s attention was on *Cumberland*, the *Beaufort* and *Raleigh* took up station on *Congress*'s stern quarters and continued to fire into her, trading fire with the shore batteries as well.

There was an extraordinary amount of iron flying around the water just off Newport News Point. The federal shore batteries, no more than a few hundred yards away, opened fire with their heavy columbiads, which *Virginia* returned with a withering accuracy even as she continued to pound the frigates. *Congress* was blasting away with everything she could bring to bear, as was *Cumberland* right up to the final moments. *Beaufort* and *Raleigh* added their firepower. Also in the mix was the Union gunboat *Zouave*, a former tugboat armed with two 30-pounder Parrott rifles.

Zouave had been serving as a picket boat at the mouth of the James River at night and a tender to *Congress* and *Cumberland* during the day. Now, as she trained her guns on the ironclad, she found herself uncomfortably between the federal fire from shore and the return fire from *Virginia* and the Confederate gunboats.

Then into that firestorm came the rest of the James River Squadron, *Patrick Henry, Jamestown*, and *Teaser*. John Tucker had anchored his squadron as far downriver as he dared, anticipating *Virginia*'s attack.

Steam was up and the anchors heaved short when *Virginia* finally made her appearance. Tucker ("that gallant officer," Buchanan called him) led his squadron downriver, steaming in line ahead, desperate to join the fight. They were still coming downriver when *Virginia* rammed *Cumberland*. The men on the gunboats cheered wildly.

Before Tucker's boats could join the battle, they had to run the gauntlet of the shore batteries. Confederates and Federals exchanged a fierce cannonade as the gunboats passed, but most of the fire on both sides went high. Still, damage was done. A shell exploded in the midst of the gun crew at *Patrick Henry*'s number-three gun, wounding two men and killing another, whose dying words, in the tradition of martial selflessness, were, "Never mind me, boys."

The nimble gunboats steamed past *Virginia* and plunged into the fight, firing fore and aft at the federal ships and shore batteries. But the Yankees, too, had more resources to bring to bear. Anchored near Fortress Monroe were two big steam frigates, both near-sisters of *Merrimack* and each one carrying more firepower than all of the Confederate vessels combined. The *Roanoke* mounted two 10-inch Dahlgren smoothbore pivot guns, twenty-eight 9-inch Dahlgrens, and fourteen 8-inch guns. *Minnesota* carried thirty-six 9-inch Dahlgrens, four 100-pound Parrott rifles, and one massive 200-pound Parrott. Also at anchor nearby was the sailing frigate *St. Lawrence*, which mounted mostly 32-pounders along with ten 8-inch guns.

The Federals, knowing *Virginia* would be out soon, kept close watch on the south side of Hampton Roads. Confederate prisoner Sharpe, onboard *Congress*, reported, "Bright lookouts are always kept up on Pig Point, Sewell Point, Elizabeth River, and all batteries. Not a boat ever escapes observation." At 1:00 P.M., one of the lookout boats made signal 551, "enemy vessel is coming."

Louis Goldsborough, flag officer of the North Atlantic Blockading Squadron, was not in Hampton Roads. Since early February he had been involved, with General Burnside, in a joint amphibious assault to clear the Confederates out of Hatteras Inlet and the Sounds of North Carolina. It was an important operation, but Welles assigned a more disgraceful motive to Goldsborough's absence. In his diary he wrote,

I have never been satisfied with the conduct of the flag-officer in
those days, who was absent in the waters of North Carolina—pur-
posely and unnecessarily absent, in my apprehension, through fear of
the Merrimac, *which he knew was completed, and ready to come*
out. He has wordy pretensions, some capacity, but no hard courage.
There is a clan of such men in the navy, who in long years of peace
have been students and acquired position, but whose real traits are
not generally understood.

It was a damning critique, and true or not, it spoke to the genuine
terror sailors had of the ironclad building in Gosport.

With Admiral Goldsborough gone, Captain John Marston, commander of *Roanoke*, was the senior officer present. He immediately signaled for *Minnesota* to get under way, and called for the tugs assigned
to his ship to come alongside. *Roanoke*'s engine was disabled with a
broken crank pin and had been for some time. Making the repairs
would have put the ship out of commission for two months. Goldsborough had opted to keep her in Hampton Roads for as long as possible,
thinking "she can still be made very useful in case of a movement on
the part of the *Merrimack*." Now was her chance.

William Radford, captain of the *Cumberland*, was onboard
Roanoke, taking part in a court of enquiry. Frantic to get back to his
ship, Radford went ashore and procured a horse, correctly assuming he
could get back to his command faster that way.

It was over 7 miles by land from Old Point Comfort to Newport News
Point, where Radford hoped for a boat to take him out to *Cumberland*,
and the captain rode hard. Thomas Selfridge claimed that as soon as
Radford arrived at Newport News, his horse keeled over and died. The
health of the horse notwithstanding, Radford arrived just in time to watch
Virginia ram and sink his ship. Stranded on the shore, he missed his
chance to participate in the most significant naval event of his lifetime.

The federal ships at Hampton Roads comprised a "noble squadron
of the lame and halt." *Minnesota* was the only substantial Union ship
that could move under her own power. At 12:45, even before the lookout boat's signal, Captain Gerson Van Brunt saw Buchanan's squadron
coming into the Roads. He ordered all hands on deck and the anchor

chain slipped. They were soon under way, making for Newport News Point, about an hour's steaming away. At one-thirty, while *Virginia* was still closing with *Congress* and *Cumberland*, *Minnesota* cleared for action. The tide was near the top of the flood and would soon be ebbing, which would force the big ship to fight the current as she tried to bring her guns to bear on *Virginia*.

Roanoke also slipped her chain and was taken in tow by the tug *Dragon*, made up alongside. Soon after, the tug *Young America* also came to *Roanoke*'s assistance, and in *Minnesota*'s wake the two boats towed the disabled frigate into battle.

The *St. Lawrence* had only arrived in Virginia two days before, and had spent her time anchored in Lynnhaven Bay, five miles east of Fortress Monroe, riding out the storm that had kept *Virginia* dockside. It was not until two o'clock that the Union gunboat *Cambridge* steamed into the bay and informed *St. Lawrence*'s captain, Hugh Purviance, of the fight that was just beginning. Purviance ordered, "All hands up anchor." By the time the sailing frigate was under way, taken in tow by *Cambridge*, the slaughter had already begun.

As *Minnesota* passed Sewell's Point, she came under fire from the Confederate batteries there. One shell did substantial damage to the mainmast, which was fished (a sort of splint) and further supported with a hawser over the masthead. *Minnesota* returned fire with her broadsides and bow pivot gun, with little effect.

Van Brunt and the men of the *Minnesota* could only look beyond the ship's bows and watch the destruction taking place off Newport News. Taking the middle, or "swash," channel, she closed to within a mile of where *Virginia* was tearing into the Union blockaders. Just moments after the ironclad rammed the *Cumberland*, *Minnesota* ran aground.

The *Minnesota* drew 23 feet, but the bottom where she touched was "soft and lumpy," and Van Brunt thought he might be able to power over the shallows. He went full ahead on the engines, but managed only to drive the ship harder into the mud. He could not back her off. The tide had begun to fall more than an hour before. *Minnesota* was stranded, a sitting duck, and *Virginia* would be coming for her next.

Congress Revisited

With *Cumberland* destroyed, Buchanan turned his attention back to *Congress*, but turning his attention was easier than turning his ship. Huge and unwieldy, *Virginia* could take as long as forty minutes and a great deal of room to reverse direction. To find sea room to make that turn, the pilots steamed *Virginia* a little ways upriver. The crew of *Congress*, thinking she was breaking off the engagement, began to cheer. They were soon disabused of their hope.

As *Virginia* began her long turn to starboard, her stern gun came to bear on *Congress* and John Taylor Wood took his first shots, raking the frigate. *Congress* was also under fire from the Confederate gunboats *Beaufort* and *Raleigh*, which continued to decimate the frigate's crew with their bow guns. All the while *Congress* fired back, and now *Minnesota*, aground a mile away, also added what firepower she could, though there was little hope of her guns effecting much from a mile away when *Cumberland*'s had proven useless at point-blank range.

To add to *Congress*'s problems, she was now on fire. The heated shot from *Virginia* had set her blazing between decks, in the sick bay, the wardroom, the main hold, and, most ominously, near the after magazine. The ship's fire company went to work fighting the blaze, while the men at the guns continued to fire at their various antagonists. As one of *Congress*'s gunners, Frederick Curtis, put it with marvelous understatement, "It was a pretty busy time aboard just then, and the men were much excited."

Lieutenant Joseph Smith, the son of Admiral Joseph Smith of the Ironclad Board, had only recently been promoted to command of the *Congress*. In fact, her old captain, Commander William Smith, was still onboard as a guest. Watching *Cumberland*'s death at the end of *Virginia*'s ram (and not knowing, of course, that the ram had been left in *Cumberland*'s side), Joe Smith knew he had to get his ship to shallower water. He slipped his anchor chain, set jib and topsails, and signaled to *Zouave* to come alongside. There was hardly a breath of wind to drive the vessel under sail.

Zouave left off her firing at *Virginia* and the gunboats and drew along *Congress*'s port side. There her captain found a hellish scene of controlled chaos and terror. The screams of wounded men were sharp over the sounds of thunder of broadsides and the roar of the flames consuming the gun deck. *Congress*'s men were so occupied with the fire and gunnery that there was no one to take the lines from the tug. Crewmen from the *Zouave* had to clamber up the frigate's side to make the towlines fast.

Lieutenant Smith ordered the tug to go ahead with her helm hard-astarboard. He was heading *Congress* for the shallow water that extended out from Newport News Point, a place where *Virginia* could not follow.

It took *Virginia* more than half an hour to turn and come back down-river, "in consequence of the shoalness of the water and the great diffi-culty of managing the ship when in or near the mud." In fact, her keel was dragging across the bottom a good part of the time, making her even more slow and unmanageable. During the maneuver, the ironclad twice had to steam past the shore batteries, which poured fire into her. But *Virginia* returned fire and got the best of it, destroying two vessels tied to a wharf.

It was three-thirty in the afternoon when *Cumberland* went under, leaving only her masts showing above the river, and *Virginia* came within range of *Congress*. The ironclad took position about 150 yards astern of the stranded ship and let go with her broadsides. *Congress* had run bow-first onto the mud. Now her stern was exposed to *Virginia*'s broadside and the only guns she could bring to bear on the enemy were the two long 32-pounders rigged as stern chasers.

Raleigh was temporarily out of the fight with a gun carriage that had become disabled, but *Beaufort* kept at it. Lieutenant Parker maintained a continuous barrage with the 32-pounder on his boat's bow, stationing himself on *Congress*'s starboard quarter. "The fire on this unfortunate ship was perfectly terrific," he recalled.

Congress was in an untenable situation, but she fought on. Gun crews worked the stern guns until they were swept away by *Virginia*'s shells, and then another crew stepped forward and worked the guns until they in turn were shot down. But the 32-pounders were useless against *Virginia*.

The fire company managed to get the fires under control, but they were not extinguished, and more fires continued to break out as *Vir-ginia* pounded the frigate. Wounded men laying on the deck of the cockpit were soaked with frigid seawater as the firefighters fought the flames. Men brought below for treatment by the surgeon were killed as shells exploded in the sick bay.

Dr. Shippen described the scene between decks. "In the wardroom and steerage the bulkheads were all knocked down by shell and axe-men, making way for the hose, forming a scene of perfect ruin and desolation. Clothing, books, glass, china, photographs, chairs, bedding, and tables were all mixed in one confused heap."

The dead and wounded were mounting. The captain of the *Zouave*, still tied alongside, recalled, "The blood was running from the *Congress* scuppers on to our deck, like water on a wash-deck morning."

A gang of cooks and wardroom servants had been organized as a powder brigade, passing cartridges from the after magazine back to the gun crews at the stern chasers. Shippen heard the gun crews calling out for powder and wondered why the flow of cartridges had stopped. He made his way into the gloom of the berth deck and found that a shell from *Virginia* had blasted right through the line of men, killing or wounding them all. Supervising the powder division, but unhurt, was Paymaster McKean Buchanan, Franklin's brother.

For forty minutes *Congress* endured the punishment, fighting back as hard as she was able. The men at the after guns "were swept away from them with great rapidity and slaughter by the terrible fire of the enemy." Finally the guns themselves were destroyed, one dismounted, the other with its muzzle knocked off. *Congress* was utterly defenseless.

Around four-twenty in the afternoon, Lieutenant Joe Smith was killed just as so many of his men had been killed, struck down by a shell fragment to his chest. Ten minutes later, first officer Austin Pendergrast learned that he was in command of *Congress*. There were no guns that would bear on the enemy, the men were being slaughtered, and the ship was on fire. The *Minnesota*, which they had hoped would come to their aid, was clearly stuck fast.

Pendergrast consulted with Commodore William Smith, who had no official role onboard but was still an experienced officer. Both men agreed the cause was hopeless. The Stars and Stripes came down, and white flags were run halfway up the mainmast and up to the peak of the spanker gaff. The firing stopped. After two hours and twenty minutes of continuous battle, the Confederate ironclad *Virginia* had won a devastating and unequivocal victory.

But the fighting was far from over.

Chapter 28

That Ship Must
Be Burned

It is a noble thing for a ship to go down fighting with her flag still flying, the gallant, last-ditch effort of a man-of-war.·

Participants on both sides of the fight lauded *Cumberland* and her men for doing so. The *Virginia*'s Lieutenant Eggleston was virtually alone in pointing out that it was "more reasonable to assume that on a rapidly sinking ship no one in the rush to save his life paused long enough to perform the quite unnecessary task of pulling down a flag upon which no enemy was firing."

Congress did not go down with flags flying. And while no one ever accused the men of *Congress* of shirking, still the surrendering ship did not come in for the same praise that the sinking one enjoyed. This despite the fact that *Congress* fought for more than two hours while *Cumberland* was destroyed about fifty minutes after the start of the battle.

Surprisingly, both ships suffered about the same number of casualties: *Cumberland* had 121 men killed out of a company of 376, while *Congress*, out of a company of 434, lost 136 men killed, wounded, or missing, the vast majority of whom were killed. It was a stunning and devastating toll.

As the battle raged off Newport News, the other Union ships were desperate to get at *Virginia*. Around the time the ironclad was making her run at *Cumberland*, *Roanoke* was passing Sewell's Point under tow. Like *Minnesota*, she traded shots with the Confederate batteries at long range, suffering no more than some torn-up rigging.

From the angle at which *Roanoke* was approaching, the men onboard could not see *Cumberland* around the tip of Newport News Point. It was around three-thirty when *Cumberland* came into view from *Roanoke*'s deck, just in time for the Roanokes to see her heel over and sink. They stood on, moving slowly thanks to *Roanoke*'s poor steerage, and could do nothing but watch as *Virginia* poured shot into the *Congress*. They watched with "mortification" as *Congress* hauled down her flag. Soon after that, *Roanoke* ran aground.

She touched at the stern, the lowest part of the ship's keel. Marston ordered the *Dragon,* with the forward towline, to turn the ship around, and this succeeded in getting her off the bottom. But now, stuck in the shallows with a falling tide, still several miles from the action, Marston did not think he could get any closer than he was. He cast off his tow and sent the tugs to aid the still-stranded *Minnesota*.

While *Roanoke* was floundering around in the shallows, *Cambridge,* with *St. Lawrence* in tow, steamed past, still heading for the fight. *Roanoke* sent her topgallant masts up and set sail, heading back for Fortress Monroe. As *Virginia*'s flag lieutenant Robert Minor described the event, putting a Confederate spin on it, she "turned tail, and, as the sailors say, 'pulled foot' for Old Point."

While sending up the masts, a lead block in the deck pulled free and clobbered seaman John McDonough on the head. It was *Roanoke*'s only casualty of the day.

Vile Treachery

With the white flags flying at *Congress*'s main and mizzen, Buchanan passed the order to stop firing. On *Virginia*'s dark, choking gun deck the cry went up, "The *Congress* has surrendered!" The lieutenants, eager for a better view than they could get through the tiny gun ports, charged for the ladders leading up to the shield deck, but Catesby Jones stopped them in midflight.

"Stand by your guns, and lieutenants, be ready to resume fighting at the word. See that your guns are well supplied with ammunition during the lull." With the Union squadron still in the offing and plenty of daylight left, Jones had no illusions that the fighting was done for the day.

Buchanan did make his way to the shield deck. The scene that greeted his unobstructed vision was greatly changed from what it had been a few hours before. Then there had been two lovely, tall frigates with bright clothing drying in the sun. Now a gray cloud of gun smoke hung over the water. Upriver, *Cumberland*'s masts jutted above the surface, the yards twisted at crazy angles, the rigging filled with men waiting to be taken up by one of the federal boats pulling out from the shore.

Congress was listing over as the tide fell and left her higher on the mud. *Zouave* had cut loose, her captain getting Pendergrast's permission to leave them. The frigate's stern was shattered, and blood ran in dark lines from her scuppers and down her sides. Threads of smoke lifted from her hatches where she was still on fire. Those among her crew who could still walk were climbing into the few undamaged boats and pulling for Newport News, unwilling to end up as prisoners of war. In fact, Buchanan did not have the resources or interest in taking any but officers prisoner.

In two hours *Virginia* had proven to the world the absolute dominance of iron cladding over wood, and Buchanan intended to further demonstrate that proof on the other ships. But he was not quite done with *Congress*.

Lieutenant William Harwar Parker had already shown that he was not a man to wait around for orders, and with the surrender of *Congress* he demonstrated that tendency again. With no word from the flag, Parker launched a boat and sent Midshipman Charles Mallory and acting volunteer Ivey Foreman over to *Congress* to take possession of the ship and bring her commanding officer back to *Beaufort*.

Mallory and Foreman drew up alongside the *Congress*, where they were greeted by a marine leveling a musket at them. Undaunted, Mallory informed the marine that he was ordered to board the vessel and was "bound to do it." The marine let the Confederates onboard unmolested.

The two young officers stepped onto *Congress*'s deck and announced that they would be taking charge of the ship, but seemed unsure of what to do after that. According to Shippen, they did nothing but "gaze about a little, and pick up one or two carbines and cutlasses, I presume as trophies." Among the weapons Mallory retrieved was Lieutenant Joseph Smith's sword, which he found with the young officer's body.

Finally they noticed that the Stars and Stripes was still flying from the gaff, so they hauled it down.

About the time that Mallory and Foreman were boarding *Congress*, Buchanan hailed *Beaufort*. When the gunboat drew up close to the ironclad, Buchanan ordered Parker to do essentially what he had begun to do without orders—go alongside *Congress* and take the Union officers prisoner. Parker was also ordered to take the wounded off of her and let the others escape to the shore. When *Congress* was clear of the living, she was to be burned.

When *Beaufort* came alongside *Congress's* port gangway, Mallory and Foremen returned to their ship and gave the flag to Parker. Mallory also showed Parker the sword he had taken. On the blade was engraved "Jos. B. Smith." Smith and Parker were old friends, classmates from the Academy who had last seen one another in Rio. Then Parker had spent the night onboard *Congress*, visiting Smith, before sailing for home in the *Merrimack*. Parker took the sword and later gave it to Commodore Tattnall, who returned it to his old shipmate, Admiral Joseph Smith.

Parker sent another of his officers aboard *Congress* to find the commanding officer and bring him back to *Beaufort*. He sent a gang of his own crew to help remove the wounded.

Beaufort's midshipman Virginius Newton stepped through the *Congress's* gangway into a scene of horror. "Confusion, death and pitiable suffering reigned supreme," he recalled, "and the horrors of war quenched the passions and enmity of months." Newton and the men of the *Beaufort* began organizing the Yankees for evacuation. One of *Congress's* crew recalled, "Some of the rebels acted like crazy men and would drive our men about like cattle. They became so abusive that one of our men, a darky, shot one of them," though no one else seemed to recall that happening.

Lieutenant Austin Pendergrast, commanding officer of the *Congress*, and Commander William Smith stepped across to *Beaufort's* hurricane deck, which was level with the frigate's main deck, to formally

surrender. Parker met them there, expecting the Union officers to surrender their swords, but Pendergrast and Smith were carrying no
sidearms. Parker told them to return to *Congress* and retrieve their
swords, which the Union officers did, to some extent. Smith returned
with his sword, but for some reason Pendergrast came back with a
navy-issue cutlass, which he presented to the much-annoyed Parker.

Parker informed Pendergrast of his orders. Pendergrast argued
against burning the ship, claiming there were too many wounded onboard, but Parker would not be moved. Even as the officers were debating the point, the *Beaufort*'s men were transferring the wounded to
the gunboat's decks. *Raleigh* drew up next to *Beaufort*, and Parker ordered her to go along *Congress*'s starboard side and begin taking
wounded off that way. Lieutenant Alexander complied.

What happened next stemmed from a tragic disagreement over
what it meant for a vessel to surrender, a misunderstanding that would
end with each side accusing the other of a cowardly disregard for the
conventions of war.

With her white flags up, *Congress*, by all rules of warfare, was out of
the fighting.

But to the Union commanders onshore, *Congress*'s white flags
meant nothing. Just because the ship had surrendered, they felt, did
not mean that the shore batteries had to cease fire.

Brigadier General Joseph Mansfield was in command of Union
troops at Newport News. Seeing the Confederate ships tied up alongside *Congress*, Mansfield ordered two infantry companies from the
nearby 20th Indiana Regiment to spread out on the beach and fire on
the enemy gunboats. A battery of two rifled guns and a rifled Dahlgren
howitzer, manned in part by fourteen sailors who had escaped from the
sinking *Cumberland,* were hidden onshore by sandbanks and trees.
These, too, opened up on *Beaufort* and *Raleigh*.

From less than 800 yards away, the Union fire was devastating. If
Mansfield had thought about it, he might have realized he had as good
a chance of hitting federal sailors as he had Confederates, and that was
exactly what happened. "At the first discharge," Parker wrote, "every
man on deck of the *Beaufort*—save Captain Smith and Lieutenant Pendergrast—was either killed or wounded." Midshipman Hutter of the

Raleigh was killed and Lieutenant Tayloe mortally wounded. Four bullets passed through William Parker's clothing. One took off his cap and his eyeglasses, another grazed his left knee.

The small arms and artillery from shore riddled the gunboats with bullet holes, making it impossible for anyone to live on the open decks. Parker told Pendergrast to order the Union troops to cease fire, but Pendergrast had no control over the infantry onshore.

Pendergrast then suggested that Parker hoist a white flag, fearing all of *Congress*'s wounded would be killed, but Parker was not about to do that. He pointed out that there *was* a white flag flying, from *Congress*, and the men onshore were firing on it.

Pendergrast and Smith asked if they could return onboard the *Congress* and help in getting the wounded onboard *Beaufort*, and Parker agreed. He was happy for the help, and he did not feel he could lock his prisoners in his cabin as he would otherwise have done, with "bullets . . . going through it like hail."

It was too hot for *Beaufort* and *Raleigh* to stay where they were. Parker blew his steam whistle, and those of his men who were still aboard *Congress* tumbled back onboard the *Beaufort*. Also onboard the gunboat were twenty-six Yankee prisoners, some wounded, some by fire from the Union infantry onshore. Pendergrast and Smith, who had given themselves up as prisoners, remained onboard *Congress*.

Lieutenant Robert Minor, onboard *Virginia*, claimed, "as Alexander backed out in the *Raleigh* he was fired on from the ports of the *Congress* . . . A dastardly, cowardly act!" Buchanan in his report also says that *Congress* fired on the gunboats.

Whether *Congress* fired or not is an important question. The fire from the shore batteries was arguably legal—the batteries had not surrendered—but if the men of *Congress* fired after the white flag was up, then it was a dastardly act indeed.

Congress gunner Frederick Curtis recalled that his gun was still loaded when the order to cease fire was passed, but, he wrote, "I pulled the lanyard and fired what proved to be the last shot ever fired on board the fated *Congress*." That was long before *Beaufort* had come alongside the ship. By the time *Congress* was supposed to have fired on the gunboats, a good number of *Congress*'s still able-bodied crew had decamped for shore. It seems unlikely that those few left behind had the means or the inclination to load and fire the great guns.

None of the Confederates who were onboard the gunboats, including William Harwar Parker, claimed that the *Congress* had fired on them. More likely, Buchanan and the others onboard *Virginia*, watching through the haze of gun smoke from 150 yards away, mistakenly believed that *Congress* as well as the shore batteries had fired.

Mistaken or not, Franklin Buchanan was in a rage, his legendary hot temper now near the melting point. But there was nothing much he could do until Parker returned and reported to him what had happened.

Parker, however, was more interested in fighting back. He moved away from *Congress,* trained *Beaufort*'s single gun on the Yankees onshore, and opened up on his antagonists. He could make little impression, however, on the entrenched enemy while his own vessel was riddled. "The sides and masts of the *Beaufort* looked like the top of a pepper box from the bullets, which went in one side and out the other." Virginius Newton observed, "Why every man on her decks was not slain or wounded is one of those phenomena which battles alone reveal."

Encumbered with prisoners, his fire ineffectual, Parker broke off the fight. But rather than report to Buchanan and inform the admiral that the *Congress* had not been set on fire as ordered, Parker steamed off toward Norfolk and delivered the wounded to Flag Officer French Forrest, standing by in the tug *Harmony*. This failure to report, Buchanan wrote to Mallory, "caused the delay in the destruction of the *Congress* and the wounding of a number of men."

For this, and for firing without orders at the onset of the battle, Buchanan told the secretary that Parker was "unfit for command." Mallory never acted on the suggestion, however, and Parker went on to have a distinguished career in the Confederate Navy.

"Destroy That Damned Ship!"

Buchanan, standing with a handful of his officers on *Virginia*'s shield deck, waited for the black smoke to start pouring out of *Congress*, indicating that Parker, who was steaming away, had set her on fire.

Buchanan "took it for granted" that that had been done, but as he waited and watched and no smoke came from the ship, he realized it had not.

The *Minnesota* was nearby, and the other ships were coming up. Buchanan feared that if he left *Congress* intact she would be retaken. Catesby Jones suggested another motive as well—that Buchanan feared accusations that he allowed the ship to be retaken to protect his brother. Whatever his reasoning, Buchanan had no intention of letting *Congress* escape. He turned to Robert Minor and said, "That ship must be burned."

Minor immediately volunteered to go. Miraculously, one of *Virginia's* boats was still in floating condition, and this was put in the water and manned. Buchanan ordered the *Teazer* to cover the boat as she pulled for *Congress*.

In a letter to John M. Brooke, Minor wrote, "I did not think the Yankees on shore would fire on me on my errand to the *Congress*." Why he thought that is unclear. Eggleston recalled Minor's boat having a white flag, which, if it did, would explain why he thought he should be immune from gunfire, though a white flag may not have been very appropriate for his mission to burn the *Congress*.

White flag or no, when Minor's boat got within a couple hundred yards of *Congress*, the shore batteries and infantry opened up on it. Again, the proximity of the shore made for an effective fire. Minor recalled, "The way the balls danced around my little boat and crew was lively beyond measure." The Confederates boldly pushed on through this hail of lead, even after two of the boat crew were cut down. Minor cheered them on, and then he, too, was shot, a bullet striking his ribs and glancing off, "coming out over the heart."

Minor was knocked to the bottom of the boat, and his men began to panic. Then the lieutenant pulled himself up again and, despite the blood pouring from his chest, calmed them as *Teazer* drew alongside and took them all onboard.

Tucker's James River boats had been active throughout, getting themselves into the thick of the fighting. Now *Patrick Henry* came forward, steaming into the hornet's nest of gunfire. *Patrick Henry* was partially plated with iron to ward off shot from her boilers and machinery, but the plating was inadequate to protect her from the fire she was taking from Newport News, the shore battery near *Congress*, and the long-range fire from *Minnesota*.

A shot took out a chunk of her walking beam, and another cut in two the sponge held by the sponger at the after pivot gun. Apparently in shock, the sponger cried out, his voice filled with despair, "O Lord, how is the gun to be sponged!" He was quickly handed a new sponge.

Soon the Yankees took out something more crucial than a sponge. A shot from the shore battery came through the side, hit a boiler, and carried away its steam chest, a chamber mounted on top of the boiler. Scalding steam roared out through the hole where the steam chest had been and filled the fire room and engine space. Lieutenant James Rochelle recalled, "[f]ive or six of the firemen were scalded to death; the engineers were driven upon deck, and the engine stopped working."

Patrick Henry was drifting closer to the enemy's guns, but her gun crews did not pause in their loading and firing. Tucker set a jib (*Patrick Henry* had an auxiliary sailing rig) to bring her head around. *Jamestown* steamed up and took a towline and towed her out of the line of fire, where repairs could be made.

Franklin Buchanan lost the last vestige of forbearance. He sounded the recall for *Teaser* and shouted down the hatchway to Lieutenant Eggleston, in charge of the hotshot guns, "Destroy that damned ship!" Some witnesses recall Buchanan standing defiantly on the shield deck, firing on the shore batteries with a rifle.

Whether he was shooting or not, he was certainly exposing himself with a reckless disregard for his own safety, and he paid the price. A moment later he was shot down. A minié ball from shore (or perhaps *Congress*—years later a marine was heard bragging about having fired the shot) hit Buchanan in the left thigh, grazing the femoral artery, a potentially fatal wound. He collapsed onto the grating and was carried below. As command devolved to Catesby ap Roger Jones, Buchanan ordered him "to fight her as long as the men could stand to their guns."

Buchanan was laid out in his cabin and Jones went to work. During the entire proceeding with *Congress*, *Virginia* had been exchanging a hot fire with the shore batteries, and taking fire from them and from *Minnesota*. Now she turned her guns back to *Congress*. Solid shot heated in the boiler fireboxes was placed in iron buckets, loaded in the ash hoist, and run up to the gun deck. There it was carried with tongs to the muzzles of Lieutenant Eggleston's guns, readied with well-soaked wadding packed on top of the powder cartridge. Shot after shot—heated bolts glowing red as well as exploding shells from the other guns—was poured into *Congress*.

By the federals' lights, this action was as dastardly as that of which the Confederates had accused the Union. "[W]e had a white flag flying to show we were out of action," Dr. Shippen pointed out, "and we were certainly not responsible for the action of the regiment on shore." But white flags were meaningless by that point, and the men on *Congress* did not hope to get any protection from them. They scrambled over the bows of the ship, dropping into the water or into boats that had pulled out from shore.

Frederick Curtis found a bucket lanyard hanging from a gun port and used it to climb down into the water. "The water about me was filled with men. Keeping as far from them as possible I started for the shore, nearly half a mile away."

Minnesota and *St. Lawrence*

Virginia did not spend more than fifteen minutes shelling *Congress*. "We raked her fore and aft with hot shot and shell," wrote Eggleston, who was personally firing the heated shot, "till out of pity we stopped without waiting for orders." *Congress* was now sufficiently ablaze that the fires would not be extinguished, though it is unclear if she was destroyed by that final cannonade, or if the fires were the same as had been burning since the start of the fight.

The tide had been falling for more than four hours and evening was approaching, but the *Virginia* still had other business to attend to. She backed away from *Congress* and headed for *Minnesota*. Not far away, the *St. Lawrence* was also aground, and along with *Minnesota* had been lobbing shells at the ironclad.

Virginia made the best of her way toward the Union ships, joined by *Jamestown* and *Patrick Henry*, running on her one undamaged boiler. It was around 5:00 P.M., and the pilots were not happy with the situation. The treacherous Middle Ground Shoal stood between *Virginia* and *Minnesota*, and the pilots did not dare take the ironclad any closer than a mile away. From that distance the Confederate squadron and the Union squadron exchanged fire.

The Union shells did no damage to *Virginia*, and *Virginia's* long-range fire was ineffectual. The ironclad scored one solid hit on *St. Lawrence,* which according to the official report "penetrated the starboard quarter about 4 inches above the water line, passed through the pantry of the wardroom and into the stateroom of the assistant surgeon on the port side, completely demolishing the bulkhead." The shell then bounced off an iron bar and ricocheted back into the wardroom. Luckily, it did not explode.

Virginia also scored a single hit on *Minnesota*, sending a shot through her bows, but the shallow-draft gunboats were able close to within a few hundred yards and do real damage. Taking position on the frigate's port bow and stern, where the Union guns could not reach, *Jamestown* and *Patrick Henry* ripped into the *Minnesota*, killing two men and wounding several others.

For nearly two hours the scattered ships pelted one another with iron. Finally, with darkness coming on and the tide still falling, *Virginia's* pilots could take no more. They insisted that the ship steam for Sewell's Point and moor for the night, and Jones, realizing it was the only thing to do, agreed.

Virginia was just getting under way when the tug *Young America* freed *St. Lawrence* from the mud. The Union ship lashed out one last time at the ironclad. Virginius Newton was charmed by the sight of the gunfire in the failing light. "The sight was a pretty one, and the *St. Lawrence,* in particular, at nightfall made a simultaneous discharge of her port broadside, which lit up for a moment the entire scene." Of that broadside, *St. Lawrence's* captain reported, "certainly no serious damage could have been done."

Virginia, with her consort of wooden gunboats, steamed off to Sewell's Point. The men were disappointed to have to break off the action. "We withdrew most reluctantly when further victory seemed so nearly in our grasp," Eggleston wrote. But they had achieved an extraordinary victory already, handing the U.S. navy the worst defeat it had ever suffered or would suffer until 1941.

As they were steaming across the Roads to Sewell's Point, Catesby Jones passed by Eggleston's station. "A pretty good day's work," Eggleston said to the new commanding officer.

"Yes," Jones said, "but it is not over."

He had no idea.

Chapter 29

On the Crater
of a Volcano

If any one of a thousand things had gone differently—if *Monitor*'s steering gear had worked the first time, if the ironclad board had taken one day less to make their decision, if Ericsson had been able to hire the night gangs as he wished—then perhaps the *Monitor* would have been there to stand between CSS *Virginia* and the helpless wooden ships at the mouth of the James River. But she was not.

Conversely, if anything had happened to slow the vessel up anymore—if the navy had made Ericsson replace the *Monitor*'s rudder, if the weather had held the ship in New York City for one more day, if Stimers's mishap with the guns had resulted in real damage—then *Virginia* would have certainly destroyed the *Minnesota* and possibly the *Roanoke* and *St. Lawrence* as well, or the latter two would have been forced to flee in disgrace. The Confederates would have been the undisputed owners of Hampton Roads. But that did not happen, either.

The timing of the two ships' meeting is an extraordinary bit of fate. The chance that two radically different vessels, begun six months and 300 miles apart, should arrive on the field of battle within six hours of one another, defies the imagination. It is one of the reasons that the story of the *Monitor* and *Virginia* has earned its unassailable place in American legend.

Once *Monitor* passed between Cape Henry and Cape Charles and stood in toward Hampton Roads, she was safe at last from the weather. The nightmare voyage down the coast was over, and she now faced the kind of danger for which she was built.

Every passing moment, every mile made good, brought the sounds of the battle more distinctly to the ears of the men on the ironclad's turret, and made them more desperate to get into the fight. Keeler recalled, "Oh, how we longed to be there—but our iron hull crept slowly on & the monotonous clank, clank of the engine betokened no increase of its speed." Standing into Hampton Roads, they could see the burning wreck of the *Congress*, their first real glimpse of the terrible damage *Virginia* had done.

Around 9:00 P.M. *Monitor* anchored at last, near Fortress Monroe and the frigate *Roanoke*, which had returned from its abortive attempt to get in the fight with *Virginia*.

The last order Worden had received was to bring *Monitor* to Hampton Roads, but as he was preparing to get under way, Welles changed his mind. Welles telegraphed Commodore Paulding, commander of the navy yard in New York, instructing him that *Monitor* should be sent to Washington, "anchoring below Alexandria." But *Monitor* was already under way before the telegram arrived, and had too big a lead for the boat Paulding sent after her to deliver the message. Worden did not get the new orders.

Arriving in Hampton Roads, Lieutenant Worden reported to the senior captain, John Marston onboard the *Roanoke*. Marston had also received orders to send *Monitor* to Washington, but in light of the day's events, that was clearly out of the question. Marston instead ordered Worden to steam to Newport News, where the *Minnesota* was still hard aground.* *Monitor* was to stand by her and defend her from *Virginia*, which would certainly be back in the morning to finish the work left undone.

An hour after dropping her anchor, *Monitor* raised it again and steamed across Hampton Roads to Newport News. Like a beacon, the

---

*Marston would make much of this decision, though Gustavus Fox felt he did no more than he should have done. Writing to Welles, he said, "Marston magnifies his facts . . . Whatever his orders may have been he must know that in battle the senior in command is a despot, can take life and of course suspend all orders." Fox blamed Marston for allowing his ships to become scattered far from the protection of Fortress Monroe.

burning *Congress* illuminated the water and helped the *Monitor* make her way to the grounded vessel.

Worden anchored alongside *Minnesota* and went onboard to confer with Captain Van Brunt. Since ten o'clock in the evening, when the tide had begun to flood, Van Brunt had been engaged in a desperate effort to get his ship off the bottom. The mud in which she was stuck acted like a suction cup holding her in place, and the recoil of her guns from the previous day's fighting had pushed her even farther down into it. Full flood tide would be at 2:00 A.M., and Van Brunt was getting steam tugs and hawsers ready for a big pull.

Worden asked Van Brunt if there was anything *Monitor* could do to assist, but Van Brunt said there was not. Worden then promised to stand by and render what protection he could against *Virginia*. Van Brunt thanked him. "[A]ll on board," he wrote, "felt that we had a friend that would stand by us in our hour of trial."

Van Brunt wrote those words the day after the battle, after *Monitor* had proven herself, but they might not be an honest reflection of how he felt the night *Monitor* arrived. It was understood that *Monitor* was the vessel built to confront *Virginia*, but most people had only a general idea of what she looked like. The reality of the shotproof iron battery was somewhat underwhelming. Many at Hampton Roads did not see her as being equal to her task. With only 18 inches of freeboard, *Monitor*'s hull could hardly be seen from a distance. All that was really visible of her was a 20-foot round tower carrying a paltry two guns to *Virginia*'s ten.

By herself, *Monitor* was not awe-inspiring, but compared to CSS *Virginia*, with her impenetrable casemate almost 200 feet long, and mounting five times the number of guns *Monitor* carried, the Union battery seemed absurd. Lieutenant Greene wrote, "An atmosphere of gloom pervaded the fleet, and the pygmy aspect of the new-comer did not insure confidence."

USS *Congress*

The survivors onboard *Congress* were relieved to see *Virginia* steam away, but that was not an end to their troubles. The ship was on fire, the flames working their way toward the powder magazine. The men left onboard felt as Shippen did, like "men would who were walking on the crater of a volcano on the verge of eruption."

Luckily for them, *Virginia's* fire had been concentrated on the gun and berth decks, leaving the upper works—and the ship's boats—relatively unscathed. With stuffing in the few holes made by flying shrapnel, the boats were seaworthy enough to get men ashore. Just as the sun was going down and the fire was really beginning to take hold, the crew rigged yard and stay tackles and swayed the launch and the first cutter over the side.

The boats pulled between the dying ship and the shore, depositing the men and returning for others. Finally a line was rigged from the fore yard, and the wounded in cots were lowered down into the launch. The water was gaining fast on the leaking boat, and as it approached the shore for the last time, the officers were obliged to jump out to lighten it.

In all, about thirty wounded were brought ashore, and one who was already dead: Lieutenant Joseph Smith. The officers, wanting to bring Smith's body ashore but knowing the men were in no mood to waste effort on a corpse, put the lieutenant's remains on a cot and sent it with the wounded.

As darkness gathered around Newport News, the stranded sailors from both ships were given shelter and food by the soldiers in the encampments there. Thomas Selfridge organized a unit of *Cumberland's* men to counter a rumored attack by Confederate forces overland, but most of the men were too exhausted to do much of anything. Frederick Curtis said, "I went with a part of the 5th Indiana, and shall never forget their kindness to us poor stranded sailors."

Virginia at Anchor

It was full dark when *Virginia* dropped her one remaining anchor off Sewell's Point. Like all the ships in Hampton Roads, her way was lit by the blazing *Congress*, now fully consumed by flames. Engineer H. Ashton Ramsay described the sight in the wonderful purple prose of the Victorian writer: "Every part of her was on fire at the same time, the red-tongued flames running up shrouds, masts and stays, and extending out to the yard-arms. She stood in bold relief against the black background, lighting up the Roads and reflecting her lurid lights on the bosom of the now placid and hushed waters."

Virginia was doing much better than her victim. She had suffered two killed and nineteen wounded. Two of the guns had muzzles shot off, and her ram was still in *Cumberland*'s side. Her less substantial gear had been annihilated. Lieutenant Wood recorded, "Railings, stanchions, boat-davits, everything was swept clean. The flag staff was repeatedly knocked over, and finally a boarding-pike was used." But the iron clad casemate had stood the test. She was bruised (Surgeon Dinwiddie Phillips counted 98 dents the following morning) but not broken. Her most signifcant damage was to the stem, the timber forming the bow, which was twisted in the course of having the ram wrenched off and now leaked. But she was certainly ready to fight again.

The rest of the squadron anchored near *Virginia* and the officers came onboard the flagship for a joyous reunion. With Buchanan's wounding, John Tucker assumed command of the squadron, but elected to remain with *Patrick Henry*, leaving command of the *Virginia* to Catesby Jones.

There was a great deal of work to do, getting the battle-scarred vessels ready for the next day's action. *Virginia*'s crew labored until midnight, and only then were they able to break for supper, their first meal since dinner, taken while steaming up the Elizabeth River.

Most of the men preferred the fresh air of the shield deck with its spectacular view of the *Congress*, wrapped in flames. Third Assistant Engineer E. A. Jack was "too tired to stay on deck with my shipmates and watch the revel of the flames on the burning *Congress*." He retired to his stateroom, only to find that *Virginia*'s two dead men had been stored there. "I did not care much for such room mates, it was too suggestive." Jack found other accommodations.

Most of the men slept at their guns, while others, too keyed up to sleep, remained on the shield deck and watched the "grand sight" of *Congress*, "all ablaze from her deck to her top," the spectacle enlivened every now and again when a still-loaded gun fired off into the night.

"A Grander Sight Was Never Seen"

Around midnight, Lieutenant Samuel Greene went aboard *Minnesota*, taking one of the two small boats *Monitor* carried lashed to the deck. He returned to *Monitor* sometime around 12:30 A.M. As he stepped back on the iconclad's deck, the fire sweeping over USS *Congress* reached the frigate's magazine and she exploded.

Congress went up in a massive fireball that illuminated Hampton Roads, *Monitor*, *Minnesota*, and the half-sunk hulk of the *Cumberland*. One by one her powder tanks went off in a series of blasts that sent debris spinning hundreds of feet in the air and raining down flaming wreckage on the dark water.

One observer recalled, "Pieces of burning timbers, exploding shells, huge fragments of the wreck, grenades & rockets filled the air & fell sparking & hissing in all directions." The men watched in stunned silence. The column of flame seemed to hang in the air, not a flash but a solid thing that remained for a minute or two before collapsing into the sea. The shock wave from the explosion rolled over them and seemed to lift *Monitor* from the water. A "grander sight was never seen," Greene recalled, "but it went straight to the marrow of our bones."

From the shorelines that ringed Hampton Roads, soldiers and civilians who had watched the battle now watched the doomed *Congress* burn. General Mansfield and the troops at Newport News were the closest to the flaming wreck. They had had a busy day, but despite the considerable shelling from the *Virginia* and the gunboats, they had had only one casualty, a private from the 7th New York Volunteers who was wounded by a shell from *Virginia*, resulting in the amputation of his leg.

The burning wreck cast its light over the beach, "illuminated the heavens and varied the scene by the firing of her own guns and the flight of her balls through the air," as Mansfield described it, until "her magazine exploded and a column of burning matter appeared high in the air, to be followed by the stillness of death."

Across the water, 4 miles away at the mouth of the Nansemond River, Confederate brigadier general R. E. Colston also watched the final act of the March 8 battle. "The burning frigate . . . seemed much nearer. As the flames crept up the rigging, every mast, spar and rope glittered against the dark sky in dazzling lines of fire. The masts and rigging were still standing, apparently almost intact, when . . . a monestrous sheaf of flame rose from the vessel to an immense height."

Three and a half miles away from the burning ship, the men standing on *Virginia*'s shield deck were stunned by the sight. Eggleston was officer of the deck. He witnessed "a sudden lightening up the sky, followed by a heavy explosion." Ramsay gushed, "The magazines exploded, shooting up a huge column of firebrands hundreds of feet in the air, and then the burning hull burst asunder and melted into the waters, while the calm night spread her sable mantle over Hampton Roads."

Preparing the Cheese Box for Battle

For a long time the *Monitor*'s men watched the terrible sight. Then, as the last of the burning *Congress* smoldered under a thick cloud of smoke, the crew turned to the work of readying *Monitor* for the next morning's action. She had been cleared for action at the first sound of the guns, just past Cape Henry, and she was ready to fight, but the interior of the vessel was in disarray after their horrific voyage down the coast. Greene and Worden remained on the roof of the turret, alert, not certain when *Virginia* would move again.

Sometime between two and four o'clock the tugs managed to get *Minnesota* free from the mud that was holding her. The frigate appeared to Worden and Greene to be drifting down on them. All hands were turned out for getting the vessel under way. The anchor was raised and the *Monitor* steamed out into the channel, clear of the bigger ship. The ironclad backed and filled for an hour, but the *Minnesota* was soon aground again, and with the tide ebbing she would stay that way for some time. *Monitor* returned to her anchorage near the frigate.

During the night, *Minnesota* took on one hundred 9-inch solid shot sent over from Fortress Monroe in anticipation of the morning's attack.

The dark hours passed slowly. By five-thirty in the morning there was nothing more for the crew to do but wait for the enemy. Some of the men tried to catch some sleep, the first rest they had had since their first night under way, forty-eight hours earlier. Some of the men, including the officers, remained awake, unsure when *Virginia* would make her appearance, unwilling to be caught off guard.

Dawn broke with still air and a light fog on the water, and a mass of activity around the *Minnesota*. The fleet of tugs was still standing by, waiting for a chance to pull the big ship off. The damage that the battery at Sewell's Point and the Confederate fleet had inflicted on *Minnesota* the day before was clearly visible. Her mainmast was "fished" and the bow and port side were punched through with holes, mostly from the rifled shot of the *Virginia*'s smaller consorts, which were able to close to within effective range.

As the mist burned off, the men onboard *Monitor* got their first look at the ship they were meant to destroy, the Confederate States Ship *Virginia*. She was anchored about three miles away, under the guns of Sewell's Point, and getting up steam. Now, for the first time in naval history, iron would fight iron.

Chapter 30

A Gleam
of Lightning

The battle of March 8 was like a cataclysmic explosion, with its center on Hampton Roads and shock waves that rippled out across the country. Army encampments, both Union and Confederate, occupied nearly every foot of shoreline bordering the water on which the ships fought. From noon on the 8th, the countryside was on high alert. Two days after *Virginia*'s first sortie, a soldier at Fortress Monroe wrote to his brother: "We have been under arms since Saturday afternoon. I have not slept 4 hours since then. I was up all Saturday night and nearly all last night, and have been on the move all the time."

The shoreline around Hampton Roads formed a natural amphitheater and soldiers and civilians crowded the shore to watch the action, making the two days of fighting arguably the most witnessed naval battle in history. "Upon the battlements of Fortress Monroe and the Rip-Raps great numbers of Union troops could be seen . . . ," one Southern civilian recalled, "while our own people lined the shores and crowded the ramparts at Craney Island and Lambert's Point . . . the scene was suggestive of the greatest performance ever given in the largest theater ever seen."

Another observer onboard the *Patrick Henry* commented, "During the battle the shores of the Confederate side of the Roads were lined with spectators from Norfolk and the adjacent country, and never, not even in the days of the gladiators, had an assemblage such a spectacle performed before them."

On the south side of the Roads, the witnesses cheered as the Confederates' victories mounted. Word spread quickly through an elated Confederacy. The South had just suffered a string of disasters, including the loss of Forts Henry and Donelson in the west and Roanoke Island and Elizabeth City in North Carolina. Richmond was under martial law, and McClellan was threatening to march on the capital with overwhelming force. The Confederates needed some good news, and the *Virginia* provided it. Mary Chesnut recorded in her diary, "The *Merrimack* business came like a gleam of lightning, illuminating a dark scene."

If the action of the 8th produced elation in the South, it had quite a different effect in the Union. The disinterested commander of the French man-of-war *Gassendi*, at anchor in Hampton Roads, vividly described the state of the Union forces after *Virginia* had done her damage:

> *Everything seemed desperate on the evening of the 8th, and a general panic appeared to take possession of everyone. The terrible engine of war, so often announced, had at length appeared, and in an hour at most had destroyed two of the strongest ships of the Union, silenced two powerful land batteries, and seen the rest of the naval force, which the day before blockaded the two rivers, retreat before her. Several vessels changed their anchorage, and all held themselves in readiness to stand out to sea at the first movement of the enemy. Everything was in confusion at Fort Monroe; ferryboats, gunboats, and tugboats were coming and going in all directions; drums and bugles beat and sounded with unusual spirit. Fort Monroe and the battery of the Rip Raps exchanged night signals without intermission.*

All through the fight, General Mansfield at Newport News had fired off one telegram after another to General Wool at Fortress Monroe, giving him a blow-by-blow account of the fight. But reports of the battle did not make it to Washington, D.C. The submerged telegraph cable from Fortress Monroe to Washington was not yet in place, the final connections being made even as the *Virginia* was tearing up the fleet, and the U.S. government was unaware of what was happening.

On the evening of the 8th the telegraph connection was completed. General Wool sent the first telegraph over the new wire, a message to Secretary of War Edwin Stanton informing him of the action of that day. Wool relayed the opinion of Marston and the other naval officers that both *Minnesota* and *Roanoke* would be captured. The message concluded with Wool's ominous remark, "It is thought the *Merrimack,* *Jamestown,* and *Yorktown* will pass the fort tonight."* Wool's belief that the ironclad would leave the Roads helped fuel the near-panic that Stanton would feel.

The Worst Day of the War

It was not until early the next morning, Sunday, March 9, that Gideon Welles heard about the debacle in Hampton Roads. He was at the Navy Department, going over the latest dispatches, when the assistant secretary of war appeared with a copy of Wool's telegram to Stanton. No sooner had he finished reading it than the summons arrived for Welles to report immediately to the White House.

Arriving at the White House, Welles found Lincoln, Secretary of State William Seward, Secretary of the Treasury Salmon Chase, and Secretary of War Edwin Stanton already engaged in a panicked discussion of the battle and its implications.

There was a gloom over the group of men such as Welles had never seen, the worst of any day during the war—a powerful testimony to the psychological effects of an unknown weapon such as *Virginia.* As bad as Hampton Roads was, it paled in comparison to other events both in bloodletting and potential danger to the Union. But the president and

---

Yorktown was the name of the *Patrick Henry* before she was captured by the Confederates and, like the name *Merrimack*, was still used by the Federals. The *Jamestown* was officially renamed the *Thomas Jefferson* after her capture but for some reason the new name was rarely used, even by the Confederates.

the secretaries were familiar with armies of men, knew what they could and could not do. Such was not the case with the ironclad monster.

Welles was the only one to maintain calm, in part because he felt he had to, to stem the growing panic in the room, and in part because he understood the *Virginia*'s limitations as the others did not.

The navy secretary was pelted with questions concerning what had been done and what would be done to defend against this threat, but he had few answers. Admiral Goldsborough had spent the past months preparing for *Virginia*'s attack, but now he was away in North Carolina and his preparations had proven worthless. Welles could say only that the Union's answer to *Virginia*, USS *Monitor*, was under way and expected in Hampton Roads at any moment.

The others were not mollified, least of all Stanton, who was in an absolute panic about what the *Virginia* would do next. He ran from room to room, sitting, standing, waving his arms, ranting. "The *Merrimac* will change the whole character of the war," he claimed, "she will lay all the cities on the seaboard under contrition!" He advocated recalling Burnside from Roanoke Island before *Virginia* blasted him out, and abandoning the navy base that had been established in Port Royal, South Carolina.

Stanton's panic was ludicrous but also infectious, and Lincoln was starting to worry. He joined Stanton in repeatedly looking out the window toward the Potomac to make sure the ironclad was not at that moment steaming up to them. Stanton felt that it was "not unlikely we shall have a shell or cannon-ball from one of her guns in the White House before we leave this room."

The secretary of war was not shy about blaming Welles for the catastrophe. The others were not nearly as accusatory, but naturally turned to the navy secretary for answers. All that Welles could say was that there was nothing much they could do, beyond waiting to see what happened. He did assure them, however, that the deep-draft *Virginia* could not pass over Kettle Bottom Shoals, which blocked her way into the Potomac River.

Seward, at least, was comforted by that assurance, the first relief he had felt that day. Stanton wanted to telegraph the mayors of New York and Boston and the military commanders in the sounds of North Carolina and Port Royal and warn them of the possibility of attack. Welles felt that would be spreading unnecessary panic, particularly as *Virginia* was not likely to be at all of those places at the same time. He repeated his faith that *Monitor* would deal with the ironclad.

"Stanton made some sneering inquiry about this new vessel, the *Monitor*," Welles recorded in his diary, "of which he admitted he knew little or nothing." Welles described the ship and the now-defunct plan to send her to Norfolk and destroy *Virginia* in the dry dock. Stanton asked about the armament. Welles informed him that she carried two guns. Stanton was dumbfounded that the ship built to counter the mighty *Virginia* had only two guns. "[H]is mingled look of incredulity and contempt cannot be described; and the tone in his voice, as he asked if my reliance was on that craft with her two guns, is equally indescribable."

Lincoln was too agitated to think straight, and though he valued Welles's opinion, he felt that he should consult with professional navy men. It was then around ten in the morning. The president ordered up his carriage and was driven to the Washington Navy Yard, where he collected up John Dahlgren, commandant since Franklin Buchanan went south. He also found General Montgomery Meigs and returned with both men to the White House.

Neither Dahlgren nor Meigs were much help. They knew little of the preparations that had been made to counter *Virginia*, and they were not, in Welles's opinion, of forceful enough character to allay fears. Dahlgren suggested that Lincoln discuss the matter with Welles. Rather than calm the civilians, Dahlgren's uncertainty just further inflamed their panic.

Around noon Welles returned home, but first he made a stop at St. John's Church, where he knew the elderly Commodore Joseph Smith would be attending service. There he informed Smith of what he knew, and told him the *Congress*, commanded by Smith's son, had sunk.

"The *Congress* sunk!" Smith exclaimed. The old man buttoned up his coat and regarded Welles calmly. "Then Joe is dead."

Welles assured him that was not necessarily the case, that many of the officers and crew had no doubt escaped.

"You don't know Joe as well as I do," Smith said. "He would not survive his ship." And indeed he had not.

Throughout the day, which Welles described as "among the most unpleasant and uncomfortable of my life," the secretaries continued to meet at the White House. They were joined by General George McClellan, also terribly worried about the goings-on in Hampton Roads. McClellan had spent months planning his big push on Richmond, which was to start with troops transported downriver from Washington and landed at Fortress Monroe. He had actually begun mobilizing his forces a few days earlier. If *Virginia* was master of Hampton Roads, then his plan was ruined.

That afternoon he telegraphed General Wool, giving him permission to abandon Newport News if the *Virginia* obtained "full command of the water." He ended his telegram by saying, "The performances of the *Merrimack* place a new aspect upon everything. I may very probably change my whole plan of campaign just on the eve of execution." McClellan would prove himself to be the most timid of all Union generals, and *Virginia* was just the sort of thing to frighten him into paralysis.

Despite Gideon Welles's admonition to not spread unnecessary panic, Stanton took it upon himself to telegraph the mayors of a number of Northern cities and warn them of the pending attack, and suggest they sink obstacles in the entrance to their harbors. Throughout that long afternoon, Lincoln and his cabinet continued to grapple with their mounting anxiety. But Welles was correct—there was nothing they could do but wait and see what happened.

They were not aware that even as their panic was reaching its high-water mark, the entire strategic picture had changed.

Chapter 31

"And Thus Commenced the Great Battle . . ."

On the morning of March 9, with a light fog lying over Hampton Roads, the *Monitor* crew finished their breakfast of hard bread and coffee and went to quarters. Soon the anchor was up and *Monitor* was under way again, steaming slowly past *Minnesota*. Worden hailed Captain Van Brunt, asking what he intended to do.

"If I cannot lighten my ship off I shall destroy her!" Van Brunt answered.

"I will stand by you to the last, if I can help you!" Worden assured him.

"No, sir, you cannot help me," Van Brunt replied. Perhaps Van Brunt's first sight of the *Monitor* in full daylight quashed what hope he had felt upon her arrival. Keeler wrote, "The idea of assistance or protection being offered to the huge thing by the little pigmy at her side seemed absolutely ridiculous & I have no doubt was so regarded by those on board of her, for the replies came down curt & crisp."

Around eight o'clock, the *Virginia* was under way and heading toward the *Minnesota*. *Monitor* put her helm over and steamed off to meet her.

"A Barrel-head Afloat with
a Cheese-Box on Top"

The *Virginia*'s men woke before sunrise, and though they had had an easy night of it, compared to the *Monitor*'s men, they were still weary, having slept only four or five hours. Dr. Shippen complained that "it seemed as though we had scarcely been asleep." To fortify themselves for the day's work, the men started with "two jiggers of whiskey and a hearty breakfast."

The first task of the day belonged to Dr. Shippen: getting the dead and wounded ashore. The dead put up little fuss, but some of the wounded—Lieutenant Robert Minor and Admiral Franklin Buchanan— did not want to leave the ship. Finally, "upon the urgent solicitations of the surgeons," as Buchanan phrased it, the two men "were very reluctantly taken on shore." Buchanan consented only because he feared that his cabin, in which he and Minor had been placed, would be needed for other wounded men. The two officers were brought over to the Confederate batteries at Sewell's Point, and later in the day transported by steamer to the naval hospital at Norfolk.

As the fog burned away, it revealed a strategic situation unchanged, as far as they could see. *Minnesota* was still aground, the other warships huddled under the guns of Fortress Monroe. *Virginia*'s first target was clear, their business of the day before unfinished.

Around eight o'clock *Virginia* won her anchor and steamed out toward Newport News and the grounded frigate, heading first toward the Rip Raps and then finding the channel down which *Minnesota* had approached the day before. Jones instructed *Virginia*'s pilots to get her as near to the frigate as they could, though with the tide far from its high point, that would not be terribly near.

Then, as they drew closer, they noticed something alongside *Minnesota*, something that had not been there the day before. It was like nothing anyone had ever seen before.

Monitor was not a military secret; the Confederates were as aware of her as the Federals were of *Virginia*, and many Southerners had a general sense of what she would look like. Presented at last with the genuine article, some recognized immediately what she was. Catesby

Jones claimed that the pilots had seen the Union ironclad the night before, silhouetted against the burning *Congress*, and they "were therefore
not surprised . . . to see the *Monitor* at anchor near the *Minnesota*."

Nor was John Taylor Wood surprised, but rather knew at once that the
"strange-looking craft" was Ericsson's *Monitor*, which, he said, "could not
possibly have made her appearance at a more inopportune time for us."

Others were not as certain. Lieutenant James H. Rochelle described the confusion onboard the *Patrick Henry*:

> *Close alongside of her* [Minnesota] *there lay such a craft as the eyes
> of a seaman never looked upon before—an immense shingle floating
> on the water, with a gigantic cheese box rising from its center; no
> sails, no wheels, no smokestack, no guns. What could it be? On
> board the* Patrick Henry *many were the surmises as to the strange
> craft. Some thought it a water tank sent to supply the* Minnesota
> *with water; others were of opinion that it was a floating magazine re-
> plenishing her exhausted stock of ammunition; a few visionary char-
> acters feebly intimated that it might be the* Monitor *which the
> Northern papers had been boasting about for a long time.*

Nor did everyone onboard the *Virginia* share Wood's certainty. Midshipman Harden Littlepage claimed that the *Virginia* was "taken
wholly by surprise when the strange vessel put in an appearance in
Hampton Roads." He first thought the vessel was a raft on which one of
Minnesota's boilers was being taken to shore for repairs. He must have
wondered why they were trying to repair a boiler when the ship herself
was about to be destroyed.

As *Virginia* steamed into battle, Lieutenant Davidson stood with
his head through a hole in the deck where the pipe for the galley stove
had been removed, just aft of the bow gun. He was looking ahead
through a pair of binoculars when he saw the strange low vessel under
way and determined that it was a raft onto which the *Minnesota*'s crew
were abandoning ship. It was not until a few moments later that Davidson climbed down and told the bow gunners, "By George, it is the Ericsson Battery, look out for her hot work."

Ramsay, watching *Monitor* move slowly away from the *Minnesota*'s side, felt she looked like "a barrel-head afloat with a cheese-box on top of it." But Ramsay at least would admit *Monitor* was no more odd than *Virginia*. "It must be confessed that both ships were queer-looking craft," he wrote some years after the fight, "as grotesque to the eyes of the men of '62 as they would appear to those of the present generation."

Cheese-box on a barrel head, tin can on a shingle, the Yankee cheese-box, any number of similes were used to describe the bizarre ship that had appeared in the night. As they steamed toward it, Lieutenant Jones conferred with his officers, including Lieutenant Davidson, as they stood on the open shield deck in the morning sun. What at sunrise had seemed a clear-cut course of action was now completely altered. The officers understood that the fight "must take place between ourselves and the enemy's ironclad." They now knew what she looked like. Soon they would see how she fought.

"That Is the *Merrimack*"

It was Worden's intention to fight the Confederates as far from *Minnesota*'s side as he could. Van Brunt wrote in his report that he had signaled *Monitor* to engage the enemy, but if he did, no one onboard *Monitor* saw the signal. They did not need an order from Van Brunt. They knew what they had to do.

As they closed with the *Virginia*, all hands stood nervously at their posts, save for Paymaster Keeler and the surgeon, Daniel Logue, who had no station for battle. The two men climbed out onto the deck where Captain Worden was watching the *Virginia*'s approach. The Confederates were still a long way off when a puff of smoke burst from her side and a shell whistled past overhead and smashed into the *Minnesota* astern.

"Gentlemen," Worden said, his tone more stern than Keeler had yet heard, "that is the *Merrimack*, you had better go below."

The two idlers did not need to be told twice. They scrambled up the ladder mounted on the outside of the turret and down the hatch in the turret roof, followed by Worden. As they climbed down through the turret, one of the guns' crew was hoisting the 170-pound round shot into the muzzle of their gun. "Send that with our compliments, my lads," Worden instructed them as he made his way down into the berthing deck, then forward to the pilothouse.

A strange, tense silence fell over *Monitor*, the only sound the thump of her engine and the slosh of water alongside. Forward in the tiny wheelhouse, three men stood shoulder to shoulder; Captain Worden; Peter Williams, the quartermaster, who would steer the vessel for the entire battle; and Acting Master Samuel Howard from the U.S. bark *Amanda*, who had volunteered as pilot.

In the engine room, third assistant engineer Isaac Newton stood by the throttle and reversing lever, while his department, the black gang, attended their duties. Coal passers dug shovels into heaps of black, dusty coal, and dumped it on the iron floor plates. The firemen heaved it into the voracious white-hot fires. The watertender fiddled with draft and stared at pressure gauges; oilers darted around with oil cans, lubricating the engine that turned the propeller shaft at a rate of fifty rotations per minute. There were over thirty oil cups on the main engine that need constant refilling, not to mention those on the blower engines, turret engines, and pumps.

There were nineteen men in the turret. Lieutenant Samuel Dana Greene was in command. Acting Master Louis Stodder stood by the control that would turn the turret on its axis. Alban Stimers, too, took his place in the turret.

Sixteen of *Monitor*'s brawniest seamen stood at the guns. John Stocking, boatswain's mate, and Thomas Lochrave, seamen, were designated gun captains. The rest stood by with rammers, sponges, and the tackles for running the guns out.

No one knew when the fight would start. Of all the men onboard, the only ones who could see beyond the confines of the ship were the three men in the pilothouse. The men in the turret could catch an occasional glimpse through the 1-inch space above the guns' barrel, but that was only when the pendulum port stoppers were open, which they were not.

Recognizing that he would not be able to tell how the turret was oriented in relationship to the vessel, Greene had marked the deck below the turret with white marks indicating bow, stern, port, and starboard.

The *Monitor*'s men had slept at best a few hours in two days. They had not had a hot meal in as long, the galley being directly under the worst leaking around the base of the turret, and they had made do with hard bread and cheese. During the past forty-eight hours they were nearly asphyxiated, some to the point of passing out, and twice had nearly sunk. They were inside an entirely novel vessel whose fighting qualities were completely unknown. All they had seen of her performance thus far was her sea-keeping ability, and that was not all they might desire. They were steaming into battle against an enemy that just the day before had handed the U.S. navy its worst defeat ever, whose casemate had proved invulnerable to everything fired at it.

It was a tense time. Paymaster Keeler described it beautifully:

> *A few straggling rays of light found their way from the top of the tower to the depths below which was dimly lighted by lanterns. Every one was at his post, fixed like a Statue, the most profound silence reigned—if there had been a coward heart there its throb would have been audible, so intense was the stillness . . . The suspense was awful as we waited in the dim light expecting every moment to hear the crash or our enemy's shot.*

Finally the silence was broken by the distant sound of a gun, then another. The men braced for the crash and clang of shot hitting *Monitor*'s turret, the telling blow that would indicate just how shotproof the shotproof battery was. But the gunfire was between *Virginia* and *Minnesota*. The big boys were ignoring little *Monitor*.

First Shots

Even if Worden had not been able to keep *Virginia* away from the *Minnesota*, the tricky shallows and low water would have done the job for him. Jones asked the pilots to get the ship within half a mile of *Minnesota*, but they were never able to approach closer than a mile off.

From that distance, *Virginia* opened up on *Minnesota*, the only target they could hope to hit. One shell struck the frigate's stern section, near the waterline, but did little damage. Though *Virginia* did not try to hit *Monitor* at long range (her turret made a target only 20 feet wide by 9 feet high—not very inviting at a mile's distance), there was no question that the Union ironclad was the main objective. Jones had made it clear that his intention was to "attack and ram her, and to keep vigorously at her until the contest was decided" by the defeat of one of the ships.

With *Virginia*'s first shots, *Minnesota* opened up as well. *Virginia* gunner Richard Curtis was impressed with their ability. "The *Minnesota* had fine gunners and many of her shots struck our Ship." But they could do no damage to the ironclad casemate.

More vulnerable were the *Virginia*'s consorts, the *Patrick Henry*, *Jamestown*, and *Teaser*, once again following faithfully in her wake. As soon as *Monitor* made her appearance, however, it was clear that they were not in the same league as the ships that would be fighting that morning. When *Monitor* opened fire, the "small fry" put up their helms and got out of the way. "I shall never forget," one observer wrote, "how they scurried back into the mouth of the river when the report of the first gun from the *Monitor* was heard."

The James River Squadron could do little that day but fire on *Monitor* from long range, and with no effect. Their bow guns and quick maneuverability had been deadly against the wooden ships, but against iron they were useless.

Iron Against Iron

Worden steamed right at *Virginia* with the shells of *Minnesota*'s broadside flying over *Monitor* with an "infernal howl."

The speaking trumpet that allowed communication between the pilothouse and the turret was broken, so Greene dispatched Keeler to ask Worden if he should open fire. Worden sent back word. "Tell Mr. Greene not to fire until I give the word, to be cool and deliberate, to take sure aim and not waste a shot." Worden did not want to fire until he was very close. Ironclad range.

Virginia was moving slowly against the current. Worden approached the enemy's starboard bow, coming on at nearly a right angle. And still *Virginia* ignored the "pigmy," concentrating on her duel with *Minnesota*.

When *Monitor* was nearly on top of *Virginia*, Worden turned hard to port. The two ships came side by side, bows pointed in opposite directions, only a few yards between them. Worden ordered the engine stopped, and sent word to the turret to fire. The great port stopper was hauled out of the way and the gun run out. Lieutenant Samuel Greene personally jerked the lanyard. The gun roared out, and flew inboard along its track. And as Greene wrote, "thus commenced the great battle between the *Monitor* and the *Merrimac*."

"A Pigmy to a Giant"

It was a stunning contrast to see the two vessels side by side. Captain Van Brunt of the *Minnesota* wrote in his report that *Monitor* approached *Virginia*, "completely covering my ship as far as was possible with her dimensions, and, much to my astonishment, laid herself right alongside of the *Merrimack,* and the contrast was that of a pigmy to a giant."

"The little boat looked so insignificant, as compared to the great bulk of the *Merrimac*," another observer wrote. The young John Wise recalled that the "disparity in size between the two was remarkable," and led to a general belief, shared by Union and Confederate alike, that *Virginia* would make short work of *Monitor*. It was only after some time that onlookers began to realize, as Wise said, that "the newcomer was a tough customer."

The men onboard *Virginia* were as ignorant of *Monitor*'s qualities as the onlookers, but they quickly discovered that size does not always matter. The day before, *Virginia* had fought wooden ships even more unwieldy than she was, and with drafts at least as deep, and that were essentially stationary. Now she faced a very different enemy. "In the narrow channel the *Monitor* had every advantage, for she drew only ten feet of water [actually 11' 4"] and the *Virginia* twenty-three feet," Eggleston observed.

Monitor showed her superior maneuverability early, steaming directly at *Virginia* with little concern for keeping in channels, running down her side and firing into her. *Virginia* returned fire immediately, a full broadside at point blank range. It was the same broadside that had so decimated *Congress* the day before, but now it had "no more effect, apparently, than so many pebblestones thrown by a child."

It was a little after eight o'clock in the morning, and for the first time in history two ironclad ships had fired on one another. "The battle," as Greene later wrote, "fairly began."

Chapter 32

The First Fight
of the Ironclads

Just as the first rough weather had revealed the little defects in *Monitor's* readiness for sea, so the opening of battle revealed chinks in her fighting armor.

The first problem was the pendulum port stoppers that swung down over the gun ports. They were massive affairs, intended to be as shotproof as the battery itself. Hauling them up out of the way required an entire gun crew heaving away on a block and tackle, "vastly increasing the work inside the turret."

Ericsson would later point out that the stoppers were not meant to be lowered every time the guns were reloaded. Rather, his plan was for the turret to be rotated away from the enemy while reloading, then rotated back to fire. Unfortunately, as the *Monitor* grappled with the *Virginia*, the gunners found that while the turret itself turned with the ease, the controls for turning it were nearly frozen. During the stormy passage, with all the seawater coming on board, the reversing wheel, used to rotate the turret, had rusted up.

Acting Master Louis Stodder was stationed at the reversing wheel, but he could not turn it. Stimers, "an active, muscular man," with the strength of arm that comes from years of work in an engine room, stepped in. He was just able to work the wheel, but could not control the turret with the finesse Ericsson had intended.

Happily, finesse was not much of an issue with massive ordnance firing at an enemy mere yards away. Here at last was *Monitor*'s final test—could she stand up to heavy guns at point-blank range? Worden was extremely anxious about the turret's machinery, "it having been predicted by many persons, that a heavy shot with great initial velocity striking the turret, would so derange it as to stop it working." But as *Virginia*'s shot slammed into the turret, Worden could see that it continued to rotate as freely as ever. He turned back to the fight "with renewed confidence and hope."

Inside the turret they were also finding the answers to worrisome questions. Would the turret take the pounding? Would the reverberation of shot striking the outside be unbearable to the men inside that iron drum?

When the first shots struck the turret, they left 4-inch-deep dents that the gunners found worrisome. They mentioned this to Greene, and Greene turned to Stimers.

"Did the shot come through?" Stimers asked.

"No, sir," one of the gunners replied, "it didn't come through, but it made a big dent, just look a there, sir!"

"A big dent," Stimers answered, "of course it made a big dent—that is just what we expected, but what do you care about that as long as it keeps out the shot?"

The gunners could not argue with that logic. "Oh! It's all right then of course, sir!" With that realization, the gun crews felt considerably safer in their iron hat box. The sound of the shot striking the turret was "pretty loud," as Greene recalled, "but it did not effect us any." Their fears allayed, the gunners began to speculate on how quickly they would whip the *Virginia*.

Second Day's Fight

Virginia's gun deck was crowded, jammed with men toiling at the great guns. While Ericsson had designed *Monitor* with great labor-saving devices, such as the friction slides on which the guns moved, *Virginia* was much more of the old school. A dozen men or more swarmed around each gun, swabbing, loading, heaving away at the gun tackles to run the guns out. Powder monkeys raced from the magazine to their assigned guns with cartridges of gunpowder.

Down the length of the gun deck, battle lanterns cast a weak light over the space, with a little more trickling in through the gratings above, but it was a dark place. The lieutenants stood behind their gun divisions, urging the men to load and fire as fast as they could. Were John Paul Jones to have strolled along that deck, he would have felt pretty much at home, whereas the *Monitor*'s turret would have been utterly alien to him.

While the men on the *Monitor* were just learning about their ship's shortcomings, the men of the *Virginia* were finding new ones. The Confederate ironclad was "as unmanageable as a water-logged vessel." That characteristic had been a problem from the day she was launched, but now it was worse. The ship was leaking around the bow where the ram had been wrenched off. The smokestack was riddled with holes, affecting the way air was drawn into the fireboxes in the boilers. Ramsey reported, "The draught was so poor that it was with great difficulty he could keep up steam."

If the *Monitor*'s port shutters were unwieldy, *Virginia*'s were lacking entirely, save for the four on the sides of the bow and stern, and those open ports represented one of *Virginia*'s greatest vulnerabilities. "Had a shot from the *Monitor* entered one of our port holes," Eggleston wrote, "it would have probably killed no less than fifty men, for . . . we were quite closely packed together."

Buchanan, eager for the fight, had sailed without the port shutters, and though he probably did not regret it, he came to understand their importance. *Virginia*'s second commander, Josiah Tattnall, would write to Mallory, "Commodore Buchanan . . . advised me in the most earnest and decided tones not to engage the *Monitor* without the port covers having been fitted. He stated that two of the *Virginia's* guns had been disabled, and . . . with two exceptions [Buchanan and Minor], all his loss in men was by shots through the ports." /

As *Monitor* closed with the *Virginia, Minnesota* was forgotten. The fight was ship to ship as the two vessels lay beside one another. *Virginia* pounded *Monitor* with her big guns and peppered her with small arms fire, the latter directed particularly at the pilothouse. Curtis recalled, "The quarter gunner had placed all along the side of our ship loaded Springfield rifles," for the use of anyone who had an opportunity to take a shot. At one point, with *Monitor* right alongside, Lieutenant Davidson told Curtis, "Take one of those guns and shoot the first man you see on board of that Ship!"

Curtis and a fellow gunner, Benjamin Sheriff, trained their Springfields out of a gun port, taking aim at the men in *Monitor's* turret, whom they could see clearly. Before they could squeeze off a shot, one of *Monitor's* big XI-inch Dahlgrens came running out, pointing directly at them, and the two men thought it wiser to duck behind the side of the iron casemate. By the time they looked outboard again, the *Monitor* was gone. *Virginia* was a heavyweight fighter, swinging away at her smaller, less powerful but quicker welterweight opponent.

Baptism of Fire

Monitor had no provision for using small arms, though the federals might have used them to greater effect than *Virginia's* crew, given that *Virginia* had no port shutters and the crowd on her gun deck meant a much better chance of hitting someone. But Ericsson had bet everything on the two big guns in his revolutionary turret.

The men in the turret were stripped to the waist, and soon they were as black as the black gang with powder smoke clinging to their sweating bodies. They threw themselves at the guns and the port stopper tackles, loading and firing as fast as they could, getting off a shot every seven or eight minutes.

Paymaster Keeler and the captain's clerk, Toffy, continued to relay messages between pilothouse and turret.

"Tell Mr. Greene," Worden called down to Keeler, "that I am going to bring him on our starboard beam close along side."

Keeler raced through the wardroom and across the berthing deck, a distance of about 75 feet, and called the message up through the hatch into the turret, then raced back with Greene's reply. Awkward as it was, the communications were largely effective. Unfortunately, Keeler and Toffy were landsmen and sometimes garbled the naval jargon.

Lieutenant Greene had not slept in fifty-one hours and had been on his feet for almost all of that time. But once the first gun went off, his profound exhaustion was forgotten. He personally fired every shot. With the guns pointing forward, the pilothouse was directly in the line of fire. Greene did not dare let the gun captains fire at will, for fear that in the excitement of the moment they would put a shot right through it. The positioning of that iron box was another of *Monitor*'s design flaws, though of course Ericsson did not see it as such.[*]

Greene's visibility was limited to the inch or so of open gunport above the guns' muzzle. He soon found that fighting in a revolving, smoke-filled tower with virtually no way to see out was extremely disorienting. He could not keep track of where the *Virginia* was in relation to his guns, and would send word to Worden asking, "How does the *Merrimac* bear?"

Worden would send word back, "On the starboard beam!" or "On the port quarter!" and Greene, referring to the white marks he had made on the deck, would search for his target.

The ships went at one another for hours, often actually touching and firing with mere yards separating turret from casemate. One of the men firing back at *Monitor* was Samuel Dana Greene's old naval academy roommate, Walter Butt. "Buttsy . . . was on board the *Merrimac*," Greene mused, "little did we ever think at the Academy, we should be firing 150 lbs. shot at one each other, but so goes the World."

A few days after the fight, Greene wrote to his parents, saying, "Five

---

[*]Ericsson argued that the guns could still be trained to within six degrees of the bow. That's true enough, but it's cutting it a bit close, given that an accident could mean killing the captain, pilot, and quartermaster, and destroying the ship's steering gear. Ericsson also claimed to have considered putting the pilothouse on top of the turret but did not do so purely out of expedience of construction, a tacit admission that the foredeck was not the best location. No subsequent monitor had its pilothouse on the bow.

times during the engagement we touched each other, and each time I fired a gun at her, and I will vouch the 168 lbs. penetrated her sides."

But Greene was wrong. The *Monitor's* fire did no more damage to *Virginia* than *Virginia's* did to *Monitor*. The ships were doling out punishment that would have destroyed any man-of-war in the world, including the vaunted HMS *Warrior* and the French *La Gloire*, but they could not harm one another.

The noise inside *Monitor* was horrendous. The deafening fire of her two big guns, the crash of *Virginia's* ordnance against the turret, and the scream of *Minnesota's* shells as they flew overhead all blended into a thunderous roar.

Unlike *Virginia*, *Monitor* was not designed specifically to be a ram, but Worden was not adverse to using her as such. As the ships circled one another, Worden made a run at *Virginia's* propeller, which he guessed was vulnerable. Jones turned his ship hard out of the way. *Monitor* missed by no more than a few feet.

The *Monitor's* guns became increasingly harder to aim. More and more effort was required to get the turret revolving, thanks to the rusted reversing wheel, and once it was moving it was impossible to stop with any precision. After an hour or so of fighting, the marks that Greene had made on the deck to help orient himself were obliterated. It was pointless for him to ask Worden where the *Virginia* was bearing because Greene could no longer tell how the turret was pointing in relationship to the hull.

Greene was forced to fire the guns on the fly. The port stoppers were hauled up and left up. When the guns were loaded, Stimers set the turret revolving. As *Virginia* passed by Greene's field of vision, he fired into her. Often the ships were so close he could not miss.

Seeing this, the officers onboard *Virginia* wondered how their enemy could take proper aim with the turret constantly in motion. The answer was, they could not.

Years after the fight, the *Virginia's* Lieutenant John Taylor Wood wrote a thoughtful analysis of the event, claiming that *Monitor* was well handled but her gunnery was poor. "[H]ad the fire been concentrated on any one spot, the shield would have been pierced," he wrote, or if the *Monitor's* gunners had managed to hit the waterline, it would have been the end for *Virginia*. But neither happened, largely because Greene could not aim. "But in all this," Wood concluded, "it should be borne in mind that both vessels . . . were receiving their baptism of fire."

"Like So Many Gladiators"

The men onboard *Virginia* would have been glad to know that *Monitor*'s turret was giving the Yankees trouble, because it was making life very difficult for them. Eggleston recalled,

> Her two eleven-inch guns, thoroughly protected, were really more formidable than our ten guns of from six to nine-inch calibre [sic], and pointing through open ports. We never got sight of her guns except when they were about to fire into us. Then the turret slowly turned, presenting to us its solid side, and enabling the gunners to load without danger.

William Cline echoed Eggleston's frustration. "We could hardly get aim at *Monitor*'s guns, as they were in sight only when being fired, and would disappear immediately thereafter." John Ericsson would have been delighted to hear the Confederate's complaints. They described the turret working exactly the way he had envisioned it.

Seeing that it was pointless to fire on the turret, Jones passed the word to concentrate fire on the pilothouse, the only other potential weak spot on the low, flush-decked ironclad.

The vessels whirled around, slamming iron into each other with no effect. *Monitor*'s solid shot, propelled by only 15 pounds of gunpowder, glanced off the *Virginia*'s iron sides. *Virginia*'s shells, meant for wooden ships, "burst into fragments against [*Monitor*'s] turret."

After a while, Catesby Jones had had enough of that no-win exchange. "The *Merrimack*, finding that she could make nothing of the *Monitor*," reported Van Brunt of the *Minnesota*, "turned her attention once more to me."

Throughout the battle, *Virginia*'s gunners had fired on *Minnesota* whenever the frigate was under their guns and the *Monitor* was not.

Now they focused their exploding ordnance on the wooden ship as Jones steamed toward her to close the range. A shot from *Virginia*'s bow gun ripped through *Minnesota*'s chief engineer's stateroom, "through the engineer's mess room, amidships, and burst in the boatswain's room, tearing four rooms all into one in its passage, exploding two charges of powder, which set the ship on fire."

The fire was quickly extinguished, and as the range closed, *Minnesota* poured her fire into *Virginia*, mostly 9-inch solid shot, a powerful broadside. A marine officer informed Van Brunt that at least fifty shots had struck *Virginia* "without producing any apparent effect."

Virginia's next shot struck the tugboat *Dragon*, made up alongside *Minnesota* and attempting to tow her off the bottom. The shell struck the boiler and exploded, wounding three men who later died and gutting the inside of the vessel. The explosion caused a near-panic onboard *Minnesota* until the men realized it was not their vessel that had suffered the hit.

As *Virginia* closed with the *Minnesota*, *Monitor* came on in hot pursuit, attempting to once again interpose herself between the Confederate ironclad and the vulnerable wooden ship. *Virginia* stood on, ignoring *Monitor*.

Then, steaming up the channel toward the stranded frigate, *Virginia* ran aground.

It was easy enough to do, "in spite of all the care of our pilots," as Jones put it. Indeed, *Virginia* had been touching bottom regularly since she had first steamed into Hampton Roads. But Surgeon Phillips suspected a more nefarious cause. He claimed that the pilot later confessed to Captain A. B. Fairfax, who then told Phillips, that he had purposely run the ship aground "through fear of passing through *Minnesota*'s terrible broadside."

If that unsubstantiated accusation was true, it was a pretty dumb trick, since *Virginia* had much more to fear from *Monitor* while aground than from *Minnesota* while floating. With her maneuverability, *Monitor* could position herself at a place where *Virginia*'s guns would not bear and pound away, doing what Wood feared, concentrating fire on a single spot.

Down in the *Virginia*'s engine room, at the bottom of the ship, H. Ashton Ramsay struggled to produce enough power to get the ship off the mud. White-hot coal fires burned in each of the sixteen furnaces, generating steam in the boilers. Gangs of firemen, "like so many gladiators" with devil's-claws and slice-bars—the long tools they used to work the fires—labored to keep the fires burning hot and clean.

On top of the thunder of the great guns one deck up and the crash of enemy shot against the shield, the engine room was deafening with "the cracking, roaring fires, escaping steam, and the loud, labored pulsations of the engines." The heat was undoubtedly above 100°, possibly over 110°. It compared, Ramsay thought, to "the poet's picture of the lower regions."

Virginia needed every ounce of thrust to get her off the mud. Now the poor performance of the engines and the faulty draft from the riddled smokestack were not just an inconvenience, but a threat to the existence of the ship. Ramsay began to pile combustibles on the boiler fires, not coal, but oily rags, wood, oily cotton waste, whatever the firemen could find that would burn fast and hot and raise the pressure in the boilers. Safety valves, designed to relieve excessive steam pressure that might explode the boilers, were lashed down and disabled. The *Virginia*'s huge prop churned up the mud of Hampton Roads, but the ironclad would not budge.

As *Virginia* struggled to free herself, *Monitor* came up astern and began to systematically pound away at the casemate, which, luckily, was *Virginia*'s least vulnerable area. Nor did *Monitor* concentrate fire on a single spot, due perhaps to the difficulty of controlling the turret. But as long as *Virginia* remained aground, there was the chance for *Monitor* to inflict mortal harm.

The waterline was the Achilles' heel, and it was worse now than it had been. In a day and a half of steaming *Virginia* had burned a prodigious amount of coal, and the loss of that weight, along with the cast-iron ram and one of the anchors, had raised the ship even higher out of the water. Her foredeck, which was supposed to be two feet underwater, was now nearly awash and her thinly armored waterline even more exposed.

Had Worden known of that vulnerability, he could have ended the fight right then with a few well-placed shots to the bow. But he did not, and by closing to short range, it is likely that *Monitor*'s guns could not have been depressed enough to strike *Virginia* on the waterline. Rather than delivering a *coup de grâce*, the Yankees continued to merely bounce solid shot off the Confederate's shield.

For fifteen minutes the *Virginia* struggled to free herself from the grip of the mud. Her inability to move apparently led Jones to think her propeller was damaged. Signal flags were run up whatever was serving as a flagstaff, but hanging limp on the windless morning it took the *Patrick Henry*'s signal officer a few minutes to decipher them. Finally he read, "Disabled my propeller is."

The wooden gunboats could not live long in the middle of the terrific gunfire from *Monitor* and *Minnesota*, but if *Virginia* was stranded and disabled, they had to make the attempt to tow her off. Boldly they steamed forward, ready to help or to die trying.

In the *Virginia*'s engine room, Ramsay watched the steam pressure climb to dangerous heights, to the point where he did not think that the boilers could take the pressure. "Just as we were beginning to despair," he recalled, "there was a perceptible movement, and the *Merrimac* slowly dragged herself off the shoal by main strength."

Seeing that "the sacrifice was not necessary," the wooden gunboats retreated to their former place, beyond the line of fire, and the wild slugfest of the ironclads resumed.

Chapter 33

The Reign
of Iron

Free at last from the mud, *Virginia* turned away from *Minnesota* and steamed off. "As soon as she got off she stood down the bay, the little battery [*Monitor*] chasing her with all speed," Van Brunt observed. It looked for a moment like a good sign, *Virginia* giving up the fight, but Catesby Jones had other ideas.

The acting captain climbed down from the pilothouse to inspect the gun deck. He found Lieutenant Eggleston's division at the midship guns standing around "at rest," doing nothing.

"Why are you not firing, Mr. Eggleston?" Jones asked.

"It is a waste of ammunition to fire at her," Eggleston replied. Eggleston, at that point, had spent more than two hours watching the *Virginia*'s precious shells shatter uselessly against *Monitor*'s side.

"Never mind," said Jones, "we are getting ready to ram her."

Rather than running from *Monitor*, Jones was looking for deeper water where he could maneuver his vessel. He called Ramsay into the pilothouse and explained his intentions, telling the engineer to reverse the engines when he felt the ships hit, if he did not receive any other orders from the wheelhouse (these were the same orders Ramsay received from Buchanan—either the two officers thought alike, or Ramsay misremembered). Ramsey returned to his engine room, waiting for the impact.

Reports of how long he waited vary, but it was a while. Wood recalled maneuvering for nearly an hour, Jones calling out helm commands, ringing the engine room. "Now 'Go ahead!' now 'Stop!' now 'Astern!' The ship was as unwieldy as Noah's ark."

"Our great length and draft, in a comparatively narrow channel, with but little water to spare, made us sluggish in our movements, and hard to steer and turn," Jones reported. Finally, after all the ponderous twisting and maneuvering, Jones had *Monitor* some distance ahead and under his bow. He ordered up all steam to the engines. The *Virginia* surged ahead to the extent that she could, building the momentum that Jones hoped would crush his smaller opponent.

They struck, a glancing blow to *Monitor*'s stern. At the final moment Worden had sheered off, avoiding a direct hit. Nor did *Virginia* hit with the force she was capable of delivering. Jones claimed that there was not sufficient time to gather headway, which was probably true, given how closely Worden had been dogging the *Virginia*.

But Ramsey claimed a different reason for the relative softness of the blow. "[O]ur commander was dubious about the result of a collision without our iron-shod beak, and gave the signal to reverse the engines long before we reached the *Monitor*." The day before, when *Virginia* struck *Cumberland*, Ramsey and his black gang had been knocked to the floor plates by the impact. This time they did not even feel it.

The ramming was not without effect, but not the effect Jones wanted. The blow opened up a fast leak in *Virginia*'s bow, water gushing in where before it had just been a steady trickle. If the ironclad had hit harder, or more directly, Jones would have succeeded only in committing suicide.

"They're Going to Run Us Down"

Monitor had just run up *Virginia*'s port side and was crossing her bow, putting the Confederate ironclad in a perfect position to ram, when Worden saw what was coming. He sent Keeler running back with a message: "Look out now they're going to run us down, give them both guns."

In the turret, Greene loaded the Dahlgrens and waited. The ship seemed to hold its breath. The gunfire had made no impression on *Monitor*, but a ramming might. If the ram on *Virginia's* bow was long enough and low enough to pass under the upper hull, it would easily pierce the half-inch iron plate of the lower hull. No one onboard *Monitor* knew how big the *Virginia's* ram was nor did they know that it was no longer there, lodged as it was in the *Cumberland's* sunken hull.

As the *Virginia* bore down on *Monitor*, Worden turned hard a-port, sheering off, and *Virginia* struck *Monitor* on the starboard quarter. The impact was enough to send the *Monitor's* men reeling. They looked anxiously around for the flood of water coming in from a rent hull, but *Virginia's* glancing blow had done no damage.

Just as *Virginia* struck the *Monitor*, Greene let fly with one of the guns, hitting the front of *Virginia's* casemate square at a range of about 12 yards. To his dismay, he watched the shot rebound off the casemate, doing no visible damage. It was clear to Greene—and he was no doubt correct—that if he had been allowed to load with 30 pounds of powder, rather than the 15 on which Dahlgren insisted, then such a shot would have done crippling damage to the Confederate ship.

Close Combat

Monitor continued to whirl around *Virginia*, putting her speed and agility to good use. After *Virginia's* failed ramming attempt, *Monitor* swung around the Confederates' stern and came up on her quarter, so close her bow was against *Virginia's* side. From that distance she fired both guns at the center of *Virginia's* shield, just abreast the after pivot gun.

The impact of the shot pushed in the side of casemate 2 or 3 inches. The after pivot gun's crew was hurled to the deck by the concussion, bleeding from their noses and ears. Jones, desperate for a way to get at the maddening opponent, called for boarders away. The boarding parties mustered, but before they could go, *Monitor* was gone.

Throughout the two days' fighting, with the hail of iron thrown at her, *Virginia* had a hard time keeping her flag flying. Once again, shortly after the attempted ramming, her flag was shot down. For some reason, the Union observers felt that this time it meant surrender. Once again they were disappointed. One of *Virginia*'s crew scrambled out onto the shield deck and reattached the flag to the riddled stack.

Jones was concerned about the leak in the bow. He called Ramsey up to the pilothouse. The engineer did not think the ship was taking on significant amounts of water, since he could see no water in the crank pits, the lowest part of the ship. "With the two large Worthington pumps, besides the bilge injectors, we could keep her afloat for hours, even with a ten-inch shell in her hull," he told Jones.

Heading for Shallow Water

Staring out through the narrow slits in the pilothouse, Worden saw *Virginia*'s men massing for attack. He called down, "They're going to board us, put in a round of cannister." Keeler ran the message aft, and then forward again with Greene's reply.

"Can't do it, both guns have solid shot."

"Give them to her, then," Worden ordered, but *Monitor* was out of the way before the Confederate crew could act.

By the time *Virginia* rammed *Monitor*, the fight had been going on for more than two hours with neither ship gaining an advantage. *Monitor*'s turret was running low on ammunition. In order to hoist the huge 170-pound shot up from below, the hatch in the turret had to be aligned with the hatch in the deck; thus the turret could not be turned during that operation. Because of her 11-foot draft, *Monitor* had the option of retreating to a place where her enemy could not reach her. Worden hauled off for a shallow patch of water to resupply the turret.

Getting the big shot up was a tedious operation. While that work was being done, Worden took the opportunity to visit the turret, the only time he would be there during the battle. He climbed up the ladder from the berthing deck and then, without a word to any of the men there, climbed out of one of the gun ports and onto the open deck.

Astonished, Louis Stodder called, "Why, Captain, what's the trouble?"

"I can't see well enough from the pilothouse," Worden replied, "and I wanted to get out here for the moment to take in the whole situation." For all of Ericsson's assurances about how well one could see through the view slots of the pilothouse, the field of vision was actually very limited, even after the slits had been widened. To make matters worse, the gunfire had left a great cloud of smoke hanging over the water, with little breeze to dissipate it. It's little wonder that Worden felt the need to reorient himself.

Worden also wanted to examine a place where one of *Virginia's* shells had struck the *Monitor* at the angle between her deck and side, a point he considered a potential weak spot. With bullets clanging off the deck around him, he lay down and examined the damage, relieved to find it was minimal. He stood up again, climbed back into the turret, and told the men that the *Virginia* could not sink them if she pounded *Monitor* for a month. The men cheered. It was news they wished to hear.

Worden returned to the pilothouse. *Monitor* was ready to resume the fight. Worden headed her back to the deep water where *Virginia* lived and once again they were circling one another, trading shots, doing no significant damage.

Inside the turret, Stimers, Stodder, and Peter Truscott, seaman, were leaning against the turret's side when one of *Virginia's* shots struck. The impact threw the men to the floor. Stodder and Truscott were knocked senseless, and had to be brought below. Surgeon Logue diagnosed "concussion of the brain" and treated them with stimulants and "cold effusion to the head." The concussions were not serious, but the men remained somewhat stunned for a couple of hours. These were the only injuries so far, after more than two hours of fighting.

Stimers, fortunately, only had his hand against the turret's side when the shot struck. He was thrown to the floor with the others, but was not injured, and was able to resume his duties.

For two hours after Worden resupplied the *Monitor's* turret, the ships continued to pound away at one another. Then, just around noon, *Virginia* scored the most significant hit of the day. From a distance of 30 feet, the Confederate fired a shell at the *Monitor's* pilothouse, just as

Worden was looking out through the view slits. The shell exploded, knocking the 2-inch-thick iron roof halfway off and cracking the 9-inch-thick iron "log" second from the top. The explosion drove shards of iron and paint into Worden's face and eyes. Worden staggered back, calling for the helmsman to sheer off.

Keeler was standing below the hanging deck of the pilothouse, waiting for orders for Lieutenant Greene. "[A] flash of light & a cloud of smoke filled the house. I noticed the Capt. stagger & put his hands to his eyes—I ran up to him & asked if he was hurt.

"'My eyes,' says he, 'I am blind.'"

Keeler and Logue helped Worden get down from the pilothouse; then Keeler ran aft and summoned Lieutenant Greene. Greene left Stimers in command in the turret and ran forward. He found Worden standing at the foot of the ladder to the pilothouse, supported by Logue, his face "a ghastly sight, with his eyes closed and the blood apparently rushing from every pore."

Greene, Keeler, and Logue helped Worden to his cabin and laid him on a couch. Worden apparently believed the pilothouse to be much more damaged than it was. He turned command of the *Monitor* over to Greene, saying, "Gentlemen I leave it with you, do what you think best. I cannot see, but do not mind me. Save the *Minnesota* if you can." Now both ironclads were under the command of their executive officers.

Incredibly, neither the quartermaster, Peter Williams, nor the pilot, Howard, were injured. When Worden was hit, Williams turned the ship away from *Virginia*, following the order to sheer off, the final order Worden would give as master of the *Monitor*. With no further orders, the helmsman continued to put distance between *Monitor* and her enemy.

Those men watching the fight from the *Minnesota* had no way of knowing what had happened. All they knew was that *Monitor* was leaving the field of battle.

"For some time after the rebels concentrated their whole battery upon the tower and pilot house of the *Monitor*, and soon after the latter stood down for Fortress Monroe, and we thought it probable she had exhausted her supply of ammunition or sustained some injury," wrote Van Brunt, captain of the *Minnesota*. He clearly believed that *Monitor* had given up the fight.

But he was wrong on all counts, save for his guess about the injury. *Monitor* was not steaming for Fortress Monroe, she was just steaming, with no one to tell the helmsman where to go.

Onboard *Virginia* they also watched their antagonist sail off. "The fight had continued over three hours," Jones wrote. "To us the *Monitor* appeared unharmed. We were therefore surprised to see her run off into shoal water where our great draft would not permit us to follow, and where our shell could not reach her."

For the third time, *Virginia* closed with *Minnesota,* the wooden gunboats trailing astern. The Union frigate was still solidly aground. She had expended nearly all of her ammunition, her crew was exhausted, and the *Monitor*, the only thing that stood between her and destruction, was now steaming off for parts unknown. "I then felt to the fullest extent my condition," wrote Van Brunt.

But he would not surrender the ship. He called a meeting of his officers. If there was no other option, they decided, they would destroy the *Minnesota*. The crew began to prepare the ship for that eventuality. Then, as Van Brunt was climbing up to the poop deck, he saw the one sight that he could have most wished to see: *Virginia* was going home.

Virginia Steams for Home

Catesby ap Roger Jones waited patiently for *Monitor* to return to the fight. The Union ironclad had paused once before to resupply the turret; it was reasonable to think they were doing so again. But the minutes dragged by and *Monitor* remained over the Middle Ground shoals, too far for *Virginia*'s guns and in water too shallow for *Virginia* to follow. Meanwhile, Jones felt a growing concern for the health of his own vessel.

He called his officers together. "The *Monitor* has given up the fight and run into shoal water," he informed them. "The pilots cannot take us any nearer to the *Minnesota*; the ship is leaking from the loss of her prow; the men are exhausted by being so long at their guns; the tide is ebbing so that we shall have to remain here all night unless we leave at once. I propose to return to Norfolk for repairs." Then he asked the officers' opinions.

Not surprisingly, they all agreed with Jones, save for John Taylor

Wood, who wanted to run up to Fortress Monroe and sink or drive off the Union ships there.

Jones did not take Wood's advice. Believing that *Monitor* had given up the fight and that his ship and crew were played out, Jones put his helm up and steamed for Norfolk.

Two Victors

Samuel Dana Greene estimated that about twenty minutes elapsed from the time he left the turret to when he stepped into the pilothouse to assess the damage and assume command, though it may have been longer than that. During that time, Stimers had been operating the turret and sighting the guns, but the distance between the ships had opened up considerably. *Minnesota* was keeping up a regular fire at *Virginia* and managed to inadvertently hit *Monitor* on more than one occasion.

Greene got the *Monitor* turned around and headed back into the fight, but the fight, he found, was over. *Virginia* was steaming back toward the batteries at Sewell's Point. *Monitor* fired a couple of guns at her but did not continue the pursuit. It was around 12:15 in the afternoon, and the ships had been fighting almost without letup for four hours.

Greene's orders, as he understood them, were to protect *Minnesota*. Further, with the pilothouse severely compromised, he did not care to get the *Monitor* under the big guns at Sewell's Point, which could be laid more accurately than those onboard a moving ship. If the pilothouse were destroyed, *Monitor*'s steering would be lost, and she would no doubt be taken by the Confederates.

He turned her around again and brought her alongside *Minnesota*. "The fight was over and we were victorious," Greene wrote.

It is common enough in the aftermath of battle for each side to put the most favorable spin possible on the outcome. But the *Monitor* and *Virginia* fight was different. In that instance, both sides sincerely believed that they were the victors. It was a point that would be debated for many years to come.

What was certain was that the first battle between ironclads was over. And the reign of iron had begun.

Chapter 34

"Merely Drilling the Men at the Guns"

*"Captain Ericsson, I congratulate you upon your great
success. Thousands have this day blessed you."*

—ALBAN STIMERS

Lieutenant Samuel Dana Greene, de facto captain of the U.S.S.
Monitor, ordered the quartermaster to steer for *Minnesota*.

Down below, in the captain's cabin, Surgeon Logue was treating Worden. He succeeded in removing bits of iron and paint from his eyes and applied cold compresses to the wounds. Worden suffered a minor concussion as well. Told that the *Virginia* had returned to Norfolk and *Minnesota* was saved, Worden, with true Nelsonian drama, replied, "Then I can die happy."

Greene brought the ironclad alongside the *Minnesota*, bumping through a sea of barrels full of rice, whiskey, beans, and sugar. Van Brunt had been lightening the ship, even throwing over seven of his 8-inch guns in his ongoing desperate effort to get off the mud, on which he had been stuck for nearly twenty-four hours.

Gustavus Fox, who had come down from Washington to see about moving the wooden ships before *Virginia* could attack, instead went out to *Minnesota* to watch the battle. "When I went aboard the *Minnesota*," he wrote, "she was about to be abandoned, before the engage-

ment, because she was aground and no confidence was held in the little *Monitor* then standing out to meet the *Merrimack*."

Fox chose a good location from which to witness the historic battle.
Minnesota was not a bystander but an active participant in the action.
As the ironclads fought one another, *Minnesota* continued to struggle to
get free of the mud while blasting away at *Virginia* and receiving deadly
fire in return. Van Brunt reported, "I opened upon her [*Virginia*] with
all my broadside guns and 10-inch pivot a broadside which would have
blown out of water any timber-built ship in the world."

When *Monitor* came alongside after the battle, Fox was there to
greet them. From *Minnesota*'s high side he hailed them and praised
them for behaving as gallantly as men could, and informed them that
they had fought the greatest naval battle on record. Everyone in Hampton Roads, on both sides of the fight, understood that history had been
made that day.

Monitor was immediately surrounded by small steamers and boats
from Fortress Monroe, Newport News, and the Union men-of-war anchored nearby. All wished to extend their congratulations and no doubt
to get a closer look at the "pigmy" that had just stood up to the giant and
driven it away.

Onboard the *Monitor*, the primary concern was Captain Worden.
Among the people who came out to the *Monitor* was Lieutenant Henry
Wise, Worden's old friend and academy roommate (and, coinciden-
tally, the man who put the torch to the *Merrimack*). Wise took charge of
Worden and he and Logue transferred the wounded man to the Balti-
more boat, and accompanied him to the naval hospital in Washington.
Lieutenant Greene was now the captain and would remain so for
thirty-two hours.

While Greene, Logue, and Wise were attending to Worden, the
stewards set about preparing dinner for the wardroom, which was
served at its usual time of 2:00 P.M. The visitors who came onboard expecting to see the horrible, bloody aftermath of battle were surprised to
see instead the *Merrimack*'s officer sitting at the wardroom table dining
on steak and green peas.

"Well, gentlemen," Secretary Fox, one of the visitors, said, "you don't look as if you were just through one of the greatest naval conflicts on record."

Lieutenant Greene replied, "No, sir, we haven't done much fighting, merely drilling the men at the guns a little."

The last the Union sailors had seen of *Virginia*, she was steaming away "in a sinking condition" (a common expression at the time, and one used by many of the witnesses). But no one knew to what degree she was sinking, or when she would be ready to fight again. Vigilance was high aboard the *Monitor*, and the men kept at quarters.

A Sinking Condition

Virginia was not in nearly as bad condition as many thought and, at least on the Union side, hoped. All ships leak, and so can be said to be in a "sinking condition," it is just a question of how fast they are going down. *Virginia* was taking on water around the bows, but she was not sinking in the classic sense of the word. There were, however, other, more immediate considerations.

With the weight of the ram and the expended coal gone, and her poorly clad waterline exposed, the ironclad was dangerously vulnerable. Her engines were performing badly with the stack shot up. The tide was running out fast.

Remaining in Hampton Roads meant fighting the *Monitor* again. Four hours of battle was enough to make Jones understand that it was pointless to go up against the Union ironclad without solid shot onboard, the bow repaired, and another iron ram installed. It was too dangerous to fight with the ship riding high and the port shutters not installed. The only reasonable thing to do was to return to Norfolk and make the necessary repairs.

H. Ashton Ramsay was under the impression that the ship was steaming for Sewell's Point to spend another night at anchor. From down in his engine room he could hear the sounds of cheering. He climbed up to the gun deck and then up again to the shield deck. To his surprise, they were just passing Craney Island, heading up the Elizabeth River. The ship was surrounded by boats of every size, carrying spectators who had watched the battle. The Confederate batteries on the island were lined with soldiers. They were all cheering the returning conquerors.

Later, Catesby Jones made his official report of the ship's condition:

Our loss is 2 killed and 19 wounded. The stem is twisted and the ship leaks. We have lost the prow, starboard anchor, and all the boats. The armor is somewhat damaged; the steam pipe and smoke-stack both riddled; the muzzles of two of the guns shot away. It was not easy to keep a flag flying. The flagstaffs were repeatedly shot away. The colors were hoisted to the smokestack and several times cut down from it.

Virtually all of the damage and casualties occurred on the first day of fighting. *Monitor* had inflicted almost no injury at all.

Several of the men onboard *Virginia* categorically denied that she was sinking, but they wasted no time getting her into dry dock. The very afternoon that she returned to Norfolk in triumph she was maneuvered into the dock. One witness reported, "Tugs had to be used to get her into the dry dock at the Navy Yard, the crew pumping and bailing with all their might to keep her afloat."

Despite how well she had stood up to the pounding she received, there was a great deal of work that needed doing before she was ready for further combat.

Captains Courageous

Monitor and *Virginia* both ended the battle under the command of their executive officers. It was Greene and Jones who made the decisions that broke off the fighting. They were both criticized for their decisions. Both appeared to come up short in light of their commanding officers' dash and boldness.

Looking back on the event, John Wise, who watched the fight from Sewell's Point, said, "When . . . we saw the *Merrimac* haul off and head for Norfolk, we could not credit the evidence of our own senses. 'Ah!'

we thought, 'dear old Buchanan would never have done it.'" Armchair admirals of the Confederate navy felt that Jones should have driven the *Monitor* off and destroyed the *Minnesota*. Buchanan was unstinting in his praise of Jones, but of course Buchanan was not there during the events in question.

Five months after the fight, Jones wrote to lieutenants Hunter Davidson and Charles Simms for corroboration that he had consulted with the officers about returning to Norfolk, and that all had agreed it was the right thing to do. "It has recently come to my knowledge," Jones wrote to Davidson, "that Captain Fairfax has been making ill-natured remarks in regard to the *Virginia's* not taking the *Minnesota*."

Both officers agreed with Jones's version of events, and still felt that under the circumstances there was no choice but retire. With the pilots unable or unwilling to get closer than a mile away from *Minnesota*, there was little chance of destroying her, let alone taking her. "I am satisfied myself that you did all that any other officer could or would have done," Davidson wrote back. "It was unfortunate that you could not have had some idea of how near old Van Brunt was to leaving the ship."

More than a year after the fight, Jones was still feeling defensive about his actions as *Virginia's* commanding officer. On May 6, 1863, Jones was promoted to commander for "gallant and meritorious conduct as executive and ordnance officer of the steamer *Virginia* in the action in Hampton Roads on the 8th of March, 1862, and in the action at Drewry's Bluff on the 15th of May, 1862."

Conspicuously absent, or so Jones felt, was any mention of the fight on the 9th. For that perceived slight, Jones intended to decline the promotion, but his friends urged him to reconsider. "Dear old Buchanan" suggested that since no vessel was captured or greatly damaged, Mallory might have thought it not necessary to mention the fight. John Taylor Wood, now working with Mallory, assured Jones it was a clerical error. Wood wrote to Jones, "The question is, Can you be of more use as a commander than as a lieutenant! You owe it to the country and service to accept it." In the end, Jones did.

Samuel Dana Greene was also plagued with questions about his choice to not follow *Virginia* as she steamed for Norfolk. It was no doubt a difficult choice, and Greene probably anticipated second-guessing. Just a few days after the battle, in a letter to his parents (who were probably not disposed to find fault in their hero son), Greene still felt the need to justify his actions, stressing that he had been under strict orders to defend *Minnesota* and not go chasing after the *Virginia*. "General Wool and Secretary Fox both have complimented me very highly for acting as I did, and said it was the strict military plan to follow," he added, a bit defensively.

Even years after the war, the questions persisted. In 1868 Worden wrote to Gideon Welles, who was still secretary of the navy, concerning the fact that Greene had been "annoyed by ungenerous allusions" concerning Worden's opinion of his conduct. Worden went on to summarize the fight, and to praise Greene effusively. As to whether Greene should have followed *Virginia* to Norfolk, Worden felt that "good judgement and sound discretion forbade it."

The bold move catches the public's imagination and praise, but it is not always the prudent choice, or the best choice. It was the bad luck of Jones and Greene that their captains commanded during the hours when gallant action was called for, and they were left with the much harder choice of discretion. In the light of the historical evidence, it is clear that they each made the best choice for their circumstance, a choice made much harder by their certain knowledge that it would be questioned.

"Expressions of Gratitude & Joy"

While *Monitor* stood by, the work of floating *Minnesota* continued. Around 4:00 P.M. she came off and the tugs managed to move her about half a mile before she grounded again on a falling tide. At five o'clock *Monitor* anchored nearby, taking her familiar post as guard against *Virginia*.

Lieutenant Greene was in command of the ship and felt a bit overwhelmed, with no officers to help him as "Stodder was injured and Webber [the other Acting Master] useless." He was running on adrenaline alone. "Every bone in my body ached," he wrote. "My limbs and joints were so sore that I could not stand. My nerves and muscles twitched as though electric shocks were continually passing through them and my head ached as if it would burst. I laid down and tried to sleep but I might as well have tried to fly."

At midnight, Acting Lieutenant William Flye reported on board *Monitor* to take over as first officer, relieving Greene somewhat of his responsibilities. At 2:00 A.M. *Minnesota* was floated again, and this time she stayed afloat and was towed over to Fortress Monroe.

The next morning *Monitor* raised anchor and followed her, steaming right into a storm of adulation that could hardly be believed. As she moved through the fleet, the crews of every vessel of every size cheered without respite. Lieutenant Greene reveled in the praise, deeply moved to be, at twenty-two years old, the captain of the vessel that General Wool described as having "saved Newport News, Hampton Roads, Fortress Monroe and perhaps your Northern ports."

After *Monitor* anchored near Fortress Monroe, the visitors began to arrive, including General Wool and Gustavus Fox, all of whom heaped more praise on the men of the ironclad. This adulation was particularly sweet to the *Monitor* men after the abuse their ship had endured while she was in New York, where they had to listen to "every kind of derisive epithet" applied to the experimental craft, and had their own judgment questioned for volunteering to serve on her.

The next day, Paymaster Keeler went ashore where he found the adulation even more intense, nearly overwhelming. The battle and *Monitor* were the only subjects under public discussion, and an actual officer from the ship was the center of attention. "You cannot conceive of the feeling there," he wrote to his wife, "the *Monitor* is on every one's tongue & the expressions of gratitude & joy embarrassed us they were so numerous."

As Keeler went about his business, strangers pointed him out as being from *Monitor*. One man gave Keeler permission to draw on his insurance policy, worth $2,000. A storekeeper gave the *Monitor*'s officers *carte blanche* to stay free at his lodgings when they were ashore.

Kind as the shopkeeper's offer was, it was in vain, as no one was allowed off the *Monitor* (Keeler as paymaster, responsible for securing supplies, was an exception). With *Virginia* still afloat, the *Monitor* was the only thing that stood between the Confederate ship and the destruction of the Union fleet. She was kept in constant readiness, with steam up in the boilers and a "sharp lookout kept."

The enthusiasm felt in Newport News was echoed throughout the North. Coastal cities that just the day before had envisioned themselves laid to ruins by the Confederate iron monster now saw themselves saved by the Yankee cheese-box, and jubilation ran high.

One of the few men with Union sympathies who was not altogether happy was John Ericsson, even though he shared with John Worden the bulk of the praise for her success. Stimers wrote to Ericsson from Fortress Monroe, "You can form no idea of how very grateful the thousands of people here are to you for having produced this vessel. General Wool told me he considers you the greatest man living." The New York state legislature and the Congress of the United States both passed resolutions of thanks to Ericsson.

But Ericsson was not the kind to let unbridled praise interfere with curmudgeonly disappointment. *Monitor*'s guns, he felt, had been aimed too high, and the wrought-iron shot not used. Moreover, had *Monitor* been provided with the 15-inch guns he wanted, or even allowed a full charge in her 11-inch guns, the *Virginia* would not, in his opinion, have steamed away. He was so certain of that fact that "he could not regard the actual result with pleasure wholly unalloyed."

Ericsson managed to stick his foot in his mouth when, during a speach to the New York Chamber of Commerce three days after the battle, he claimed that *Monitor*'s success was "entirely owing to the presence of a master-mind," Alban Stimers. It is no suprise that the man who told Commodore Smith that the engineer's post on the ship was the most important office aboard should feel that way. But with John Worden now elevated to hero status, it was a heresy to give Stimers the credit for the fight.

Despite his complaints, Ericsson might have been somewhat mollified when, on March 14, the United States navy made the final payment for *Monitor*, $68,750, which represented the 25 percent held in reserve. Her fight with the *Virginia* was regarded as a sufficient test of her abilities.

Monitor's Third Captain

On the evening of March 10, the day after the fight, *Monitor's* temporary captain reported onboard. It was Lieutenant Thomas Selfridge of the *Cumberland,* who had participated in the first day's fight with *Virginia* and been so conspicuous at the abandonment of the Norfolk Naval Yard. Selfridge had been appointed by Gustavus Fox on the recommendation of Captain Marston.

On the first day of the fighting, after abandoning *Cumberland*, Selfridge and his men had gone ashore at Newport News. Seeing that *Cumberland's* flag was still flying at the gaff, Selfridge called for volunteers to man a skiff and go get it, to save it from falling into Confederate hands. He managed to accomplish this, and on returning to shore was taken to the headquarters of a New York Zouave regiment. Selfridge's clothes were soaked, and he was offered a new set—a Zouave uniform, with its ostentatious loose-fitting red trousers, short blue jacket, and red kepi.

Selfridge was still wearing that outlandish outfit when he went aboard *Monitor*. "I went below," he wrote, "and could scarcely contain my amusement at the surprise of the *Monitor's* officers, upon seeing a Zouave back down the narrow hatchway and announce himself as their new commander."

Fox sent with Selfridge a note to Greene explaining why he had decided not to leave Greene in command, due mostly to his "extreme youth" (though at 26, Selfridge was not much older or more experienced than Greene, and there was no fit line officer anywhere who knew more about *Monitor* than did her first lieutenant). Greene was surprised and disappointed by the change, but he understood Fox's thinking, giving *Monitor's* extraordinary importance.

Selfridge began immediately to familiarize himself with the ship, thinking it likely that they would be fighting *Virginia* again very soon. The next day they steamed down to the mouth of the Elizabeth River, hoping *Virginia* would come out, but there was no sign of the Confederate ship. Once again, the critical question in Hampton Roads was, "When will *Merrimack* be out?"

It would take more than a month for that question to be answered.

Chapter 35

The Last
Meeting

I f the Union was eager to know when *Virginia* would be out of dry
dock, Stephen Mallory was desperate. Eleven days after the battle
he wrote to French Forrest, saying, "The work of getting the *Virginia* . . . ready for sea at the earliest possible moment is the most important duty that could devolve upon a naval officer at this time, and
yet this Department is ignorant of what progress is being made . . . I
am not advised that a day's work has been done upon the *Virginia* since
she went into dock."

It was the old problem of matériel and manpower, though this time
Mallory had anticipated the problems and urged their timely solution.
Within an hour of hearing via telegraph that *Virginia* was engaged with
the Union fleet, Mallory placed an order with Tredegar for eight tons of
rolled iron to repair the anticipated damage. On March 10 he
telegraphed French Forrest that the iron was on the way.

Inspecting the dry-docked ironclad's casemate, John Luke Porter
found ninety-seven indentations where shot had struck the iron. Only
six plates in the upper course of iron were broken, none in the lower.
The wood backing was intact. *Virginia*'s shield had proven itself.

Mallory's iron order would have been excessive were it not for the
fact that *Virginia* desperately needed an armor belt around the water-

line. Twenty days after the battle, Porter had repaired all of the damaged iron on the casemate and mounted a new, bigger ram on the bow. He began to install the deeper, longer armor belt around the waterline. Mallory fired off nearly daily telegraphs to French Forrest and then Captain Sidney Smith Lee, Robert E. Lee's older brother, who took command of the yard in late March, urged them to hurry the *Virginia* along. Gangs of workmen labored on the ironclad seven days a week, sometimes around the clock.

Mallory had high hopes for the *Virginia*, and he was not alone. The Southerners' enthusiasm for their ironclad was every bit as fevered and unrealistic as the enthusiasm the North had for *Monitor*, the result of both sides believing they had scored a victory.

In fact, the enthusiasm in the South ran even higher than that in the North. The Confederacy had the success of March 8 to enjoy, when *Virginia* doled out such destruction to the Union navy. In late February, the Richmond diarist John B. Jones wrote, "We must run the career of disasters allotted us, and await the turning of the tide." To many Southerners, *Virginia* marked the beginning of the flood.

William Harwar Parker recalled that in Norfolk "the whole city was alive with joy and excitement. Nothing was talked of but the *Merrimac* and what she had accomplished. As to what she would do in the future, no limit was set to her powers."

The Southern press jumped on the bandwagon, publishing the "wildest speculations" as to what the *Virginia* would do next. The public ate it up. Parker recalled trying to explain to various civilians that the *Virginia* could not singlehandedly take Fortress Monroe, she could not steam to New York and level the city, she could not capture Washington. But it was no use.

Stephen Mallory was gripped with the same excessive optimism. On March 17 he wrote to Franklin Buchanan, recovering in the hospital, asking about the possibility of an attack on New York City.* "Such

---

*This letter is dated March 7 in the *Official Records of the Navy* and has been cited by historians as evidence Mallory was making big plans even before the battle. Buchanan's reply clearly refers to it as the "confidential communication of the 17th inst." It was only after *Virginia*'s command performance that Mallory grew bold enough to suggest an attack on New York City.

an event would eclipse all the glories of the combats of the sea . . . ," he claimed, hyperbolically, "and would strike a blow from which the enemy could never recover. Peace would inevitably follow."

While Mallory's hopes were a bit high, he was right in thinking an attack on New York City would have profound effects. But Buchanan dashed cold water on the secretary. "The *Virginia* is yet an experiment and by no means invincible," he explained to Mallory. It was possible, he wrote, that *Virginia* would survive a shelling from Fortress Monroe and the Rip Raps, but then she would be at sea, with a green crew, "a third of whom have never been seasick," and it was likely she would founder if she met even a very heavy swell.

Monitor would hound her the whole way to New York, and if she made it, it would be nearly impossible to get her over the bar into the harbor, and once there she would be bottled in. "I consider the *Virginia* to be the most important protection to the safety of Norfolk and her services can be made very valuable in this neighborhood," he concluded. Buchanan understood his ship's limitations, and made certain Mallory did as well. But Mallory was still desperate to get *Virginia* to sea.

"A Very Easy and Very Lazy Time"

Two days after Selfridge took command of *Monitor*, he was replaced by Lieutenant William Jeffers, Commodore Smith's original first choice, whose previous command experience made him much more qualified for the post than Selfridge. Jeffers was thirty-eight years old, pudgy, with a thin beard and mustache. He was fourth in the first class to graduate from the naval academy and had been active in much of the fighting that had taken place in North Carolina.

Jeffers entertained the wardroom with stories of his adventures, and the officers and crew liked him initially, an attitude that would change as the summer wore on. Life onboard was not nearly as pleasant under the new captain as it had been under Worden. Just two weeks after Jeffers took command, Paymaster William Keeler was describing him as "a rigid disciplinarian, of quick and imperious temper," though he always respected Jeffers' abilities as a fighting captain.

From the pinnacle of danger and action, which had been the *Monitor*'s lot since leaving New York, she now sunk into a deep idleness. So important was she considered to the protection of Fortress Monroe and Newport News, and to McClellan's ability to launch his peninsular campaign, that no one, from Lincoln on down, dared risk her in a fight, unless it was to defend against another attack by *Virginia*. *Monitor* was all the Union had to protect Hampton Roads. Nearly all the wooden men-of-war, having proven themselves useless against *Virginia*, were sent away.

Flag Officer Louis Goldsborough returned to Hampton Roads immediately after the battle and assumed command there. He wrote to Fox, "My determination is to act as strong on the defensive as possible, & to incur no avoidable risk on the part of *Monitor* beyond expending her energies carefully & guardedly against the *Merrimac* when the time arrives." *Monitor* remained on high alert. Several times word came that *Virginia* was on her way out again. The men raced to quarters and the ironclad steamed out to do battle, only to discover it was a false alarm.

On the return from one such outing, the *Monitor* passed close by the Union batteries at Newport News. Thousands of soldiers lined the shore and cheered as the ironclad steamed by, shouting so that their voices blended into one great roar. Each regiment had their band play some patriotic air, and the soldiers called out "You're the boys!" and "Bully for you!" and "Iron sides and iron hearts!" to the appreciative crew.

But for the men onboard *Monitor*, such excitement was now the exception. Mostly it was lying at anchor and waiting. George Geer summed up the change from frantic activity to somnambulance when he wrote to his wife, "We are having very easy and very lazy times laying here waiting for the *Merimack*. I have been in the Navy one month to day and I have worked so hard the time has sliped by very quick but know [now] we are doing nothing and it commences to drag very slow."

Paymaster Keeler complained to his wife that his weight had gone from 145 to 157. It is unlikely that anyone in the army, slogging up the Peninsula or marching along the Western Rivers, was having a problem with excessive weight gain.

The chief occupation of the *Monitor*'s crew was playing tour guide to the hundreds of officers, politicians, and sundry important individuals, along with their wives and daughters, who flocked to see the most famous ship in the world. "We are perfectly over-run with visitors," Keeler wrote. Chief Engineer Isaac Newton lamented to a friend about the "intolerable nuisance" of having to explain the workings of the ship again and again. The visitors were so numerous, wrote Newton, that he could "hardly afford to be polite to anything less than a Brigadier Gen[eral]."

While *Monitor* waited, she was made ready for another battle. Ammunition for the 11-inch Dahlgrens and small arms such as Enfield rifles with sword bayonets were brought onboard, the latter to defend against boarders, which were considered one of the foremost threats to the shotproof battery. Such antipersonnel weapons as steam hoses on deck and "liquid fire," consisting of ignited petroleum, naphtha, and benzine, were also discussed.

Stimers set about strengthening the pilothouse, which had proven so vulnerable. Borrowing an idea from *Virginia*'s casemate design, he built angled oak walls around the square structure so that it looked like a pyramid with the top cut off. Over the oak he put three layers of inch-thick wrought-iron plate smeared with tallow and black lead to help the enemy's shots glance off.

Jeffers was eager to get at the enemy. No doubt it was awkward for him to be in command of the ship and men who had shared such danger and rendered such great service when he himself had not been there. He asked for permission to steam to Norfolk and attack *Virginia* in dry dock. Had he been allowed to, he might have accomplished the very thing that Welles had first envisioned for *Monitor*, but he was refused. Geer wrote, "We all know what he [Jeffers] wants: he wants to distinguish himself with her as Captain Worden don, and he is afraid Worden will be back before he has a chance—and we are all praying for him to come soon."

The men of the *Monitor* missed their captain. While Worden was recuperating, they wrote as a crew to him, a wonderful expression of lower-deck sentiment.

DEAR SIR These few lines is from your own Crew of the Monitor *with there Kindest Love to you there Honered Captain Hoping to God that they will have the pleasure of Welcoming you Back to us*

*again Soon for we are all Ready able and willing to meet Death or
any thing else only gives us Back our own Captain again Dear Captain we have got your Pilot house fixed and all Ready for you when
you get well again and we all Sincerely hope that soon we will have
the pleasure of welcoming you Back to it again (for since you left us
we have had no pleasure on Board of the* Monitor *we once was
happy on Board of our little* Monitor *But since we Lost you we have
Lost our all that was Dear to us Still)*

The men were concerned about *Monitor*'s reputation, since papers
in Norfolk had called them cowards and said they were not willing to
fight. The men expressed hope and confidence that Worden would re-
join them. The letter concluded with, "We Remain untill Death your
Affectionate Crew, the Monitors Boys."

Worden would never again command the *Monitor*, but during the
little more than seven weeks that he did command her, he earned the
thanks of a nation. In New York a collection was taken up to raise
funds for a monetary reward for his performance. More than one hundred businesses and individuals donated over seven thousand dollars to
the fund. Donating one hundred dollars each were John Griswold,
John Winslow, and John Ericsson.

Abraham Lincoln instructed Welles to give Worden official thanks,
but he did more than that. Hearing that Worden was in Washington,
Lincoln said to his cabinet, "There will be no further business today. I
am going to see the brave fellow."

Lincoln took his carriage to the home in which Worden was recovering. The lieutenant accompanying Lincoln led him to the upstairs
bedroom and announced him to Worden. "Jack, here is the President,
come to see you."

"You do me great honor," Worden told the president.

At first Lincoln was too overcome with emotion to reply. Finally,
with tears in his eyes, he said, "It is not so. It is you who honor me and
your country, and I will promote you."

Worden also gave Lincoln his professional opinion of *Monitor*'s vulnerabilities, warning that she might easily be captured by boarders who
could drive wedges under the turret to prevent it from turning, then

pouring water down through the air intakes. "I have just seen Lieut. Worden . . . ," Lincoln wrote to Welles. "He is decidedly of the opinion [*Monitor*] should not go sky-larking up to Norfolk." Welles relayed the instructions to Fox in Hampton Roads.

General McClellan, always quick to overestimate a threat, was trying to determine if *Monitor* was sufficient bulwark against *Virginia*. "Can I rely on the *Monitor* to keep the *Merrimack* in check, so that I can make Fort Monroe a base of operations?" he telegraphed Fox. "Please answer at once."

Fox's answer was a definite maybe. "The *Monitor* is more than a match for the *Merrimack*, but she might be disabled in the next encounter. I can not advise so great dependence upon her."

But as March dragged on and *Virginia* remained in her dry dock, McClellan began to make the plans for his peninsular campaign, with Fortress Monroe as the base of operations.

In deference to the *Virginia* he decided to use the York River, rather than the James, for his waterborne supply line. Most people, including Fox and Welles, did not think *Virginia* could get out of Hampton Roads and into the York, whereas it was thought that *Virginia* might possibly ascend the James, or at least blockade the mouth.

By early April the peninsular campaign was under way. McClellan was moving slowly, very slowly, up the York Peninsula and there was little for *Monitor* to do but listen to the distant sound of his guns.

At the same time, Fox and Goldsborough were making plans to defeat *Virginia* when she came out again. *Virginia*'s invulnerability to shot had convinced them that ramming was the only means by which she would be destroyed. They began to charter fast steamers for the express purpose of running the ironclad down.

"Destroy the *Merrimack* by running her down is what I want all to do," Admiral Goldsborough wrote to the commanders of all the ships in the Roads. They were to "strike her with the greatest possible velocity at right angles to her side or stern," even if doing so meant the destruction of the ramming vessel as well. (The crew of one of the steamers, the *Illinois*, upon hearing the mission for which their ship had been chartered, flatly refused to participate. A new and more daring crew was found.)

Virginia's Second Voyage

A short while after the battle of March 9, a letter was sent to Stephen Mallory, written by Lieutenant Charles Simms and signed by one of *Virginia's* paymasters and three of her lieutenants, requesting that Catesby Jones be retained as commander of *Virginia* until Buchanan had recovered.

But Buchanan's wound, if no longer life-threatening, was still serious, and he would not be going to sea for a while. Jones did not have the seniority for command of *Virginia*, particularly as Buchanan was not strictly speaking her captain, but rather admiral of the James River Squadron. Jones would not be superceding Old Buck.

Instead, on March 21, Mallory assigned a friend and contemporary of Buchanan's, the sixty-six-year-old Josiah Tattnall, "Old Tat," to "command the naval forces in the waters of Virginia." With his flagship still in dry dock, Tattnall met with his officers, consulted with Buchanan, watched the slow progress of *Virginia's* refit, and planned what he would do when he met the enemy.

By April 4, *Virginia* was floating again, though her refit was not complete. Engineer H. Ashton Ramsay reported that on the first day of dockside trials, the engines broke down after turning for only a few hours. He informed Tattnall that he could "not insure their working any length of time consecutively," not a hopeful sign. But Mallory was already urging action. "Do not hesitate or wait for orders," he wrote to Tattnall, "but strike when, how, and where your judgment may dictate."

The tedious work of cutting, shaping, drilling, and installing the heavy iron plates continued night and day as Mallory continued to push for completion of the ship. Also high on Mallory's urgency list was another ironclad building at Norfolk, modeled along the lines of *Virginia*, but smaller. She would eventually become the CSS *Richmond*, built from the keel up to John Luke Porter's design.

By April 10, the bad weather that had been whipping Norfolk cleared up, revealing a sunny and windless day. The *Virginia* was still not completed, but she was close enough. In all, Porter had put on an additional 440 iron plates, extending the shield down 3½ feet from the bottom edge of the eaves and 160 feet aft from the bow. Despite Buchanan's warning not to go into battle without port shutters, not all of them had been fitted. A new and improved ram was securely bolted to the bow and the shot locker was stocked with solid wrought-iron shot. Tattnall reported to Mallory that *Virginia* was "in better condition than when she first left the yard." That was certainly true, though the additional iron also gave her an additional foot of draft, which she could hardly afford.

Just as Goldsborough and company were gathering rams to destroy the *Virginia*, Mallory, Tattnall, and the officers in the James River Squadron were coming up with ideas for destroying *Monitor*. It is interesting that neither side considered using their ironclad to destroy the other's ironclad—that had not worked so well the first time—but each hoped rather to use their ironclad to lure the other to a place where it might be attacked by other means.

The Confederates' plan was simple: while Tattnall in *Virginia* engaged the *Monitor*, the gunboats, under the command of John Tucker, would race up alongside the Union ironclad and boarders would leap onto her deck. Each of the gunboats' thirty-man crews were divided into three squads. Squad One was to throw ignited bottles of turpentine down the air intakes, smokestack, and any other opening in the deck. Squad Two was to drive wedges under the turret to stop it from turning. Squad Three was to cover the smokestack, pilothouse, and any other opening with wet sailcloth.

It was, in essence, exactly the tactic Worden had warned Lincoln about. Nor was it much of a secret. Keeler wrote to his wife that she should not believe "the thousand & one rumors you hear & will hear of our being taken by driving wedges under our turret . . ."

Virginius Newton, part of *Beaufort*'s crew, called the plan for taking *Monitor* a "forlorn hope." "I . . . would have made my will but that I had no property," he recalled. It was thought that if four gunboats made a run at *Monitor*, one at least would get through.

As *Virginia* was undergoing repairs, the gunboats' crews trained for their part, practicing the boarding maneuvers over and over again. It was not an unreasonable plan, if they could get close enough to the Union ironclad.

In the evening hours of April 10, *Virginia* steamed up the Elizabeth River for the second time, with the gunboats of the James River Squadron as well as the tug *Harmony* in company. She anchored for the night off Craney Island and was under way early the next morning. Once again, hundreds of small boats swarmed like waterbugs to cheer the ironclad and witness what everyone thought would be the rematch of the century.

Once again, the Union ships scattered as *Virginia* cleared Craney Island and made her ponderous way into Hampton Roads. A British and a French man-of-war, anchored in the Roads, weighed their anchors and slowly withdrew toward Newport News, where they expected to watch the action out of the line of fire.

Up toward Fortress Monroe and the Rip Raps lay the *Monitor*, *Minnesota*, and a smattering of Union navy vessels, including the rams intended to run the *Virginia* down. Around 7:30 A.M., seeing the *Virginia* emerge from the thin haze that hung over Sewell's Point and move into the Roads, Goldsborough fired a shot from the flagship *Minnesota*, a signal for all vessels to clear for action. The crews scrambled to ready their ships for battle. Once more, the long-anticipated moment was here. *Virginia* was coming out.

Soon steam was up in boilers, guns loaded and run out, gun crews at their stations, powder monkeys ready to race between the magazine and the guns. *Monitor* had a full head of steam and her turret was keyed up for action, but she remained on her anchor. She was not about to steam out to meet the "Big Thing," as they had nicknamed *Virginia*, and her consorts, whose intentions they had divined. But if *Virginia* made an attack on the fleet, she was ready to fight.

Tattnall gave his men a short speech, much as Buchanan had done, finishing with the words, "Now you go to your station, and I'll go to mine." Old Tat's station was an armchair placed on the shield deck where the elderly flag officer could sit and wait for the action to commence.

Virginia steamed slowly up toward Newport News, waiting for *Monitor* to attack. Tattnall had no intention of steaming toward Fortress Monroe and the crossfire of heavy guns offered by the fortress and Fort Wool on the Rip Raps, at least not until the port shutters were fitted. What was more, the Confederates had intelligence, possibly provided by one of the French naval officers, that the enemy had placed "torpedoes"—what are now called mines—in the channel between Fortress Monroe and the Rip Raps to keep *Virginia* out of Chesapeake Bay. It was not true, but the possibility was enough to encourage *Virginia* to keep her distance.

For hours the two ships eyed one another, each waiting for the other to attack, each increasingly certain the other was afraid of coming to grips once more. In frustration, Tattnall dispatched the *Jamestown* and *Raleigh* to capture three small merchant vessels that had been run aground on seeing *Virginia* enter the roads, their crews escaping overland. The Confederates took possession of the ships and towed them off, with the Union flags inverted under the Confederate flag, taunting the *Monitor*.

The Union ships watched it happen but did not budge. Onboard *Monitor*, William Keeler felt his "blood boil with indignation to see it done with such impunity." But Keeler understood the strategic situation, and knew that *Monitor*, at least, had no business racing off to protect three small ships. To the hundreds of Union people who watched from the shore, the "apparent apathy" of the U.S. navy "excited surprise and indignation."

The Northern press was less sympathetic with Goldsborough's strategy. "The public are very justly indignant at the conduct of our navy in Hampton Roads last Friday," the *New York Herald* wrote. "The wretched imbecility of the management of the Navy Department has paralyzed the best sailors and the best navy in the world."

All day *Virginia* steamed back and forth, with most of her men standing on the shield deck rather than sweltering on the gun deck below. They received waves of greeting from the French man-of-war as they passed. Finally, as the sun was setting, *Virginia* took a long-range shot at *Monitor*. The Union ship *Naugatuck* replied with a few shots from its rifled Parrott gun, neither side hitting anything. *Virginia* steamed to Sewell's Point and took up a mooring that had been set specifically for her. The day was over. It had been a nonevent.

For the next several days, *Virginia* continued to prowl Hampton Roads in sight of the Union fleet, with neither side willing to enter into the other's arena. "Had the *Merrimack* engaged the *Monitor*," Goldsborough reported, "which she might have done, I was quite prepared, with several vessels, to avail myself of a favorable moment and run her down."

Likewise, the crews of the gunboats were ready with their wedges and wet sails. But the two vessels never came to grips. Finally, *Virginia* returned to the Gosport Naval Shipyard for water and the never-ending repairs to her engines.

Virginius Newton missed his chance to board *Monitor* and smother her with wet sailcloth. "I have never yet decided," he wrote, "whether they of the *Monitor* or we of the gunboats were the more fortunate that our purpose was not put to the experiment."

Chapter 36

The End of
the Progenitors

Of the forty-five days that Josiah Tattnall commanded CSS *Virginia*, only thirteen were not spent in dry dock or undergoing some kind of repair. In total, she made five trips from Norfolk to Hampton Roads. On two of them the engines failed. Tattnall found himself "mortified beyond measure by the frequent suggestions, not only from unofficial but from high official sources, of important services to be performed by the *Virginia*, founded on the most exaggerated ideas of her qualities."

After her foray against the Union fleet in the second week of April, she once again returned to Portsmouth for another long spell in the shipyard.

Life was miserable onboard the ship. Every time there was a movement among the Union ships, the order rang out, "Clear away for action!," one false alarm after another. Steam was always maintained in the boilers, adding to the repressive atmosphere below, where "no ray of light ever penetrated." The only place to walk was on the shield deck, the foredeck, or the half-submerged fantail, which gave watchers from shore the impression that the *Virginia's* men were walking on water.

As the spring of 1862 wore on, the Confederacy was feeling the Anaconda's squeeze on several fronts. George McClellan's one-hundred-thousand-man army was dug in around Yorktown, held back by a Confederate force less than half its strength and by "Little Mac"'s excessive caution. From the west came news of the bloody fight at Shiloh, the first taste for North and South of the kind of carnage the future held. Island No. 10 on the Mississippi had fallen to the Union, and David Dixon Porter and David Farragut were preparing to move on New Orleans. And things were not looking too good for Norfolk, either.

As McClellan advanced up the peninsula, the threat increased that Norfolk would be cut off from the rest of the Confederacy. "The abandonment of the peninsula will, of course, involve the loss of all our batteries on the north shore of James River," Confederate general Joseph E. Johnston wrote to Tattnall. "The effect of this upon our holding Norfolk and our ships you will readily perceive." Johnston hoped that *Virginia* could steam past Fortress Monroe and attack McClellan's shipping in the York River. But Tattnall did not care to tangle with the forts, certainly not without port shutters in place, and his ship was needed elsewhere. In truth, it was needed everywhere.

It was not until May 3 that General Magruder abandoned Yorktown, just as McClellan was ready to start bombarding the Confederate works. On that day, Mallory wrote to Sidney Lee, telling him that due to Johnston's retreat up the peninsula, Norfolk would soon be abandoned. "The *Richmond* must be launched at once," he wrote, and he gave additional orders to begin dismantling everything that could be saved and shipping it off to safer locations within the Confederacy. "All ordnance stores and guns which can not be saved must be destroyed." The writing was on the wall. It was time to go.

Commander in Chief

It was more than a month after McClellan began operations on the peninsula that his troops entered Yorktown, about 17 miles from where they had started. Lincoln was growing impatient.

Lincoln was also growing impatient with Admiral Goldsborough and the lack of progress being made on the naval front. On May 5 the president took passage in the steamer *Miami* from Washington to

Hampton Roads, accompanied by Secretary of War Stanton and Secretary of the Treasury Salmon Chase. Exercising his prerogative as commander in chief, Lincoln began to issue orders.

The president had heard from McClellan that the batteries on Jamestown Island, once commanded by Catesby Jones, had been abandoned, and he ordered Goldsborough to send gunboats up the James River to reconnoiter. Goldsborough sent three boats, including Bushnell's *Galena*, which had recently arrived in Hampton Roads. The boats engaged Confederate batteries along the shoreline, as well as the Confederate gunboats *Yorktown* and *Jamestown*.

On May 8, James Byers, master of the tug *John B. White* and a Union man living in Norfolk, ran his boat past the Confederate batteries and steamed to Fortress Monroe. There he met Lincoln and the cabinet members and informed them that the Confederates were preparing to abandon the city. Lincoln wanted General Wool to occupy Norfolk, and the president took it upon himself to find a place to land troops. Pulling alongside *Monitor*, Lincoln personally ordered Lieutenant Jeffers to steam to Sewell's Point and shell the Confederate batteries, to determine whether or not they had been abandoned. Jeffers finished coaling and got under way.

On Lincoln's orders, Goldsborough also dispatched the *Susquehanna, San Jacinto, Dacotah*, and *Seminole* "for the purpose of shelling Sewell's Point battery, to draw out, if possible, the rebel steamer *Merrimack* into a position where she could be attacked simultaneously by the large steamers." Around twelve-thirty the squadron opened fire on the batteries at Sewell's Point, setting a barracks on fire. The few Confederates left in the partially abandoned works returned fire, but the Union observers were able to determine that "the number of men now stationed there" was "comparatively quite limited."

As the squadron poured fire into the Confederate batteries, the sound of the heavy ordnance echoed around Hampton Roads.

It was heard quite distinctly in Norfolk, and the men of the *Virginia* moved to respond.

"Their Gallant Ship Was No More"

CSS *Virginia* was considered the backbone of Confederate defenses in Hampton Roads and Norfolk, and Josiah Tattnall was being pulled in many directions.

On May 4, Mallory had telegraphed the admiral with orders to position *Virginia* at the mouth of the James River to prevent Union gunboats from ascending.

On hearing this, General Benjamin Huger, commanding Confederate forces in Norfolk, insisted that *Virginia* remain where she was. He claimed that if she left, he would be forced to abandon his batteries on Craney Island and Sewell's Point.

Tattnall could not disregard orders from the secretary of the navy, so Huger telegraphed Mallory asking that *Virginia* be retained in Norfolk. Mallory altered his orders to Tattnall, instructing him to "endeavor to afford protection to Norfolk as well as the James River," a tall and ambiguous order. Tattnall and Huger made arrangements for Huger to notify the admiral in case Norfolk was to be abandoned, so that Tattnall could take the best course of action open to him.

On May 8, a meeting was arranged between Tattnall, Huger, and Commodore George Hollins, sent by Mallory, and the other officers involved to decide on the best use for the ironclad. Before the meeting commenced, the sound of gunfire came rolling down from Sewell's Point. Old Tatt ordered the *Virginia* under way.

As usual, *Virginia* was swarming with yard workers. At the sound of the guns downriver, they were ordered to stop what they were doing and help turn the ship, which was lying with her bow upstream. This was no easy task, but finally it was accomplished and the ironclad was under way. Once again, *Virginia* steamed down the Elizabeth River, hoping to do battle.

As *Virginia* passed Craney Island and turned toward Sewell's Point, Tattnall sighted the Union squadron firing on the Confederate batteries. Closer toward Fortress Monroe he could see the *Vanderbilt*, a ram intended for running *Virginia* down, and the *Minnesota*, getting under way and heading for the action. Tattnall believed that a fight was inevitable.

But once again, it was not to be. As *Virginia* steamed toward the Union vessels, they abruptly ceased fire and headed off toward Fortress Monroe. What looked like ignominious flight to the Confederates was in fact the Union battle plan. "The *Monitor* had orders to fall back into fair channel way, and only to engage her [*Virginia*] seriously in such a position that this ship [*Minnesota*], together with the merchant vessels intended for the purpose, could run her down," Goldsborough reported to Lincoln.

But Tattnall in *Virginia* was not going to wander into that trap. Though Goldsborough also told Lincoln, "The *Merrimack* . . . was even more cautious than ever" and she "could have engaged her [*Monitor*] without difficulty had she been so disposed," Tattnall recalled it differently. In his report he claimed that he chased *Monitor* toward Old Point Comfort "until the shells from the Rip Raps passed over" *Virginia*.

Tattnall would go no farther into the lions' den than that. He called to Catesby Jones, "Mr. Jones, fire a gun to windward [an age-old naval sign of contempt] and take the ship back to her buoy." *Virginia* returned to Sewell's Point and picked up her mooring. Once again, both ships believed the other had declined battle, when in fact both had been willing to fight, just not on the other's terms.

Leaving *Virginia* at Sewell's Point, Tattnall returned to Norfolk to attend the meeting that had been preempted by the Union gunfire. Resuming their meeting on May 9, he and the other officers unanimously agreed that the ironclad "should continue for the present to protect Norfolk, and thus afford time to remove the public property." Everyone knew that the Yankees were closing in.

The next day, May 10, Tattnall was back aboard *Virginia*, still moored off Sewell's Point, when they noticed something odd: the Confederate flag was no longer flying above the batteries onshore. A boat was sent to reconnoiter and found the works deserted.

Tattnall had an agreement with Huger that the general would inform him when Norfolk was being abandoned, but Old Tatt had heard nothing. He sent his flag lieutenant, John Pembroke Jones, over to Craney Island, where the flag was still flying, to find out how things lay.

Jones returned with the news that the enemy had landed on the Chesapeake Bay side and was marching on Norfolk, that the batteries on
Sewell's Point had been abandoned, and that the Confederate troops
were retreating.

John P. Jones was then sent upriver to Norfolk to find General
Huger and Captain Lee and ask their intentions. But Jones could find
neither officer. Huger and his soldiers had taken to the trains and left
the city. Lee had supervised the destruction of everything in the shipyard that could not be moved, then he and the navy men had followed
behind Huger. The federals were in Norfolk, and the city was lost.

It was about 7:00 P.M. when Jones returned to the flagship and reported. On the way back down the river he had discovered that now the
batteries on Craney Island were also deserted. From *Virginia*'s shield
deck they could see great clouds of black smoke lifting up from the
burning Gosport Naval Shipyard. Just a little over a year since the retreating federals had set the shipyard on fire, the retreating Confederates had done the same.

The information Jones brought left Tattnall with the decision of a
lifetime. He had two choices: try to get the *Virginia* to Richmond or go
out in a blaze of glory.

If he tried to attack the Union fleet near Fortress Monroe, it was
likely that the enemy's ships would simply sail away and *Virginia* would
never catch them, subjecting herself to the deadly fire of Fortress Monroe
for nothing. The *Monitor* and the other Union ships, including the rams,
would swarm around the clumsy ironclad "as rabbits around a sloth."

If he could get to Richmond, however, the ship would be a great asset to the defense of the capital.

Virginia's chief pilot, Parrish, and his assistant, Wright, had assured
Tattnall many times that they could get the ironclad to within 40 miles
of Richmond if her draft could be reduced to 18 feet. Tattnall had already asked John Luke Porter how far the ironclad could be raised if all
weight that could be removed was, and Porter told him she could come
up to a draft of 17 feet. Tattnall consulted with his officers. The choice
seemed clear: the ship should go to Richmond.

All hands were called on deck and in the presence of Parrish and
the other pilots Tattnall explained the situation. He told them that it

would take an extraordinary effort for them to get the ship lightened in time for them to steam past Newport News in the dark. The crew replied with three cheers and fell to work.

Tattnall was unwell and went to bed as the work proceeded. The crew labored like demons, throwing off all the kentledge on the bow and fantail, tossing all spare stores overboard, all the water and food, everything that could be moved, save for the powder and shot. According to H. Ashton Ramsay, 600 tons of coal were dumped overboard.

The men worked for five hours and managed to reduce *Virginia's* draft by 3 feet when the pilots approached Catesby Jones with a little surprise. Eighteen feet of draft, it turned out, would only get *Virginia* up to Richmond in a steady easterly wind, which would tend to raise the level of the James. The wind, however, was westerly, and had been for days, which reduced the amount of water in the river. The direction of the wind and its effect on the river were no surprise to Parrish and the rest. They had just failed to mention it.

Tattnall was woken and informed of this and he was, understandably, furious. With *Virginia's* armor belt now well out of the water, there was no question of her going into battle. The ballast was at the bottom of the river. Nor could she be lowered again by flooding the bilges. The amount of water that would have been required to lower her that far would have submerged the fireboxes in the boilers. *Virginia* was doomed, and the pilots had sealed her fate.

"It will be asked what motives the pilots could have had to deceive me," Tattnall wrote in his report. "The only imaginable one is that they wished to avoid going into battle." Once again, *Virginia* was kept from fighting by her pilots. Once again, the suspected motive was cowardice.

Tattnall's options were reduced to none. There was nothing to do but destroy the *Virginia* before she fell into Union hands, and the lieutenants Jones agreed. The men were called to "splice the main brace" one last time, then *Virginia* slipped her mooring and steamed up the Elizabeth River to a place just south of Craney Island, where she was run aground as close to shore as she could get.

The crew were issued small arms and two days' provisions and ferried ashore in a chaotic and disorganized abandonment of the ship. *Virginia* had only two boats, and it was a matter of three hours' pulling back and forth before they were all landed. Harden Littlepage emptied the clothes from his knapsack to make room for *Virginia*'s tattered battle flags.

Trains of gunpowder and cotton waste were laid along the deck, just as had been done when the ship bore the name *Merrimack*. Catesby Jones and John Taylor Wood were the last to leave. They set the tinder on fire and took to the remaining boat, reaching shore just as the the first light in the east was beginning to appear.

The crew began the weary march overland to Suffolk. Tattnall, who was too weak to even stand on the shield deck for long, was in no condition to march, so the men found a cart in which to pull him, giving him a position in the center of the column. A picket guard was sent out on the Portsmouth side of the column, from where an attack was expected.

By 5:00 A.M. the column was several miles from the river when the fire reached the 18 tons of gunpowder in *Virginia*'s magazine. The earth shook and the sky lit up in a great flash of light and a roar that was heard all over the waters in which CSS *Virginia* had spent her entire life. "The slow match, the magazine," wrote H. Ashton Ramsay, "and that last deep, low, sullen, mournful boom told our people, now far away on the march, that their gallant ship was no more."

Chapter 37

Cape Hatteras

From the deck of the *Minnesota*, Admiral Goldsborough witnessed the explosion somewhere near Norfolk, and he knew either the *Virginia* or the batteries on Craney Island had been destroyed. He soon learned which it was.

Onboard the *Monitor*, the distant sound of an explosion in the early morning hours brought the crew scrambling up on deck. Like Goldsborough, they had a notion of what the explosion was. A few hours later they learned that it was the death knell of their adversary.

The men of the *Monitor* were disappointed by the news. They had, as Keeler put it, "looked upon the 'Big Thing' as our exclusive game." They felt certain she "would die game rather than fall by her own hands." No doubt the men of the *Virginia* would have wished the same.

Goldsborough dispatched the ubiquitous Lieutenant Thomas Selfridge to Sewell's Point to ascertain that it had been abandoned. Selfridge landed on the point and hoisted the U.S. flag over the works. Certain now that Norfolk was free of Confederate forces, Goldsborough sent a squad of ships, led by *Monitor*, up the Elizabeth River to Norfolk. Following in *Monitor*'s wake was a small steamer bearing Goldsborough and President Lincoln, eager to see Norfolk once again in Union hands.

On May 11, Ericsson's ironclad finally arrived at the place she was built to go—the Gosport Naval Shipyard. But now *Virginia*, her *raison d'être*, was in a thousand pieces at the bottom of the river.

Monitor was next sent up the James River to do battle with Confederate gun emplacements and the smattering of gunboats between Hampton Roads and Richmond. On May 15, *Monitor* and *Galena* carried on a four-hour duel with a Confederate battery at Drewry's Bluff, just eight miles downstream from Richmond.

Among the Confederate gunners were some of the *Virginia*'s crews, who had been sent down from Richmond two days before under the command of Catesby Jones. There had been no battery there before— the men, "exposed to constant rain, in bottomless mud and without shelter, on scant provisions," struggled to mount the guns that would hold the Yankees at bay. Exhausted, they still managed to fight the Union gunboats, doling out lethal punishment to the ill-conceived *Galena*. The flag under which they fought was the one that Littlepage had saved from the doomed *Virginia*.

Monitor spent the summer in relative idleness, patrolling the James River as McClellan carried on his astoundingly inept campaign against the new commander of the Army of Northern Virginia, Robert E. Lee. As the months passed, *Monitor*'s crew grew restless and miserable, and her machinery cried out for repair. Isaac Newton pointed out in a letter to a friend that at no time since February 20 had the fires under *Monitor*'s boilers been extinguished. The iron hull became infested with flies, which drove the men to madness.

In August the *Monitor*'s men were relieved of one irritation when Lieutenant Jeffers was replaced as captain. "I can assure you," Keeler wrote his wife, "we parted from him without many regrets. He is a person of a good deal of scientific attainment, but brutal, selfish & ambitious." Commander Thomas Stevens, who took his place, was much better liked.

A little more than a month later, Stevens was replaced by *Monitor*'s final captain, John P. Bankhead. Bankhead was a South Carolinian who had remained loyal to the Union. Not too far from the *Monitor*'s area of operation, his cousin, General John Bankhead Magruder, fought for the Confederacy.

In early October the ship was finally sent to the Washington Naval Yard for desperately needed repairs. In the eight months since her battle with *Virginia*, her fame had not diminished, and she proved a huge draw. When the yard was opened to the public, literally thousands of people flocked to see the ironclad. Entire regiments were marched to the dockyard for sightseeing duty. The carriages lining the dock made "a perfect jam." For the short time between her arrival and her being handed over to the yard workers, she was the most popular sight in Washington, D.C.

Monitor underwent a few changes in the yard. Telescopic smoke pipes, which Ericsson had first used on *Princeton*, replaced the inadequate square stacks. The air intakes also received taller pipes, and cranes and davits were installed to handle the larger boats provided. A breastwork of boiler iron about 5 feet high was installed around the top of the turret, allowing men to stand there with some protection from small arms fire. Stanchions were installed around the perimeter of the deck through which a rope railing was strung, making that open space a little less treacherous.

The dents from the hard knocks *Monitor* had taken were covered over with iron patches and each labeled as to where the shot had come from; *"Merrimac," "Minnesota," "Merrimac's Prow,"* and so on. The two big guns were engraved with the names of the two men who had emerged as the great heroes of the battle, one with $\frac{\text{Monitor \& Merrimac}}{\text{Worden}}$ and the other with $\frac{\text{Monitor \& Merrimac}}{\text{Ericsson}}$.

After her much-needed yard period, *Monitor* returned to the Hampton Roads area, where she remained for the next few months. It was not until December that some meaningful work was found for her, as part of a naval operation against Wilmington, North Carolina, where her shallow draft and heavy ordnance could be put to good use.

The enlisted men celebrated Christmas in Hampton Roads with a full day's work and what was intended to be a lavish Christmas dinner. George Geer wrote to his wife, "We had every thing to make a splendid Dinner in your hands, but our Saylor Cook made very bad work cooking to suit me." The other sailors, however, apparently not as discriminating as Geer, enjoyed their Christmas dinner, which cost the princely sum of "about $100 each."

The officers as usual fared better than the men, enjoying a three hour dinner of soups, fish, oysters, and "Meats enough to start a Chatham Street eating house."

A series of storms had lashed the coast as *Monitor* and her consorts made ready for their voyage South. Lieutenant Greene ordered Geer to see about securing the hatches so they were watertight. Geer sealed them with red lead putty and made 1-inch-thick rubber gaskets to go around the portholes. No one who had been aboard the ship on her maiden voyage would forget the way water had poured through openings all over the ship and nearly sunk them.

Once again, Ericsson's idea for a watertight joint between turret and hull, with the turret resting on a bronze ring, was not trusted by the tradition-bound sailors. Oakum was the thing that had kept water out of hulls from time immemorial, and Bankhead figured it would work again. He ordered oakum put in around the base of the tower, just as plaited rope had been used before. It would prove every bit as effective.

Around two o'clock on the afternoon of December 29, *Monitor* steamed out of Hampton Roads on her second open-water voyage. As with the first, she was under tow, trailing 300 feet astern of the side-wheeler *Rhode Island*, at the end of an 11-inch and a 15-inch hawser. The ships were bound for Beaufort, North Carolina, by way of Cape Hatteras, the "Graveyard of the Atlantic."

Around the time that *Monitor* was putting out to sea, Captain John Worden, recovered from his wounds, arrived in Hampton Roads with his new command, the turreted ironclad *Montauk*. There is no record to indicate whether or not Worden caught one last glimpse of his beloved *Monitor* as she steamed off to sea.

The weather was fine and clear as *Monitor* began her voyage, with only a light wind from the southwest, shades of her first ocean voyage, and thanks to that first experience, the men were not so sure of her sea-keeping abilities. According to Keeler, the officers felt certain that a severe gale would send her to the bottom. They would have transferred their personal effects to the *Rhode Island*, but they did not wish to panic the crew by displaying such a lack of confidence in the ship.

All through the night and into the next morning the weather continued fine. Sunrise found the *Monitor* and her consort well out to sea, with conditions so easy that the men spent most of their time on the open deck to escape the close atmosphere below.

As the afternoon wore on, the clouds began to build in the south and west, increasing until the sun was blotted out by the cold, wet, gray blanket. The swell started to build, setting in from the southward, the direction *Monitor* was heading, so she found herself punching bowfirst into the rising sea.

By midafternoon no one was taking their leisure on the open deck. The seas were crashing over it, sweeping along the ship's length, making it impossible to remain there. The top of the turret was the only safe open-air spot on board, and there the men amused themselves watching "two or three large sharks" that had begun trailing the *Monitor.*

The wind and sea continued to rise as the crew was called to dinner, but still the ship was taking on only a little water through the pilothouse and around the base of the turret, just enough to make it fairly miserable below. Still, dinner in the wardroom was cheerful, the officers optimistic about their chances for glory in the upcoming action.

When Keeler made his way up to the turret after dinner, the ship was just off Cape Hatteras, "the Cape Horn of our Atlantic Coast." Hatteras was and is one of the most treacherous places on the eastern seaboard, having claimed countless ships since mariners first ranged along the American coast. "The wind was blowing violently," Keeler recalled, "the heavy seas rolled over our bows dashing against the pilot house & surging aft, would strike the solid turret with a force to make it tremble." For the men who had been on the maiden voyage, which was most of the crew, it must have been like a recurring nightmare.

Except for one important difference. On the maiden voyage, *Monitor* was taking the waves on the beam, rolling hard but not suffering from the shock loading that comes with slamming into the sea. This time she was being towed straight into the waves. As the seas rolled under her, her bow would lift, then come smashing down into the trough between the waves. "[E]very time she rased on a sea," Geer wrote, "she would come down very heavy on her over-reaching sides and her bottom would shiver like a leaf, and I made up my mind she would not stand that long before her bottom would give away."

The ship's pumps had been working nonstop and had kept up with the inflow of water, but now word came up from the engine room that the water was gaining. The large Worthington pump was started, and for a while it was able to keep up with the inflow, but soon that, too, was overwhelmed.

Where was the water coming from? There were the usual suspects—the view slits of the pilothouse, the hawse pipe, and particularly the bottom of the turret. Geer, who had helped caulk the turret, noted, "under the Tower the Captain had Oakum put, but did not put any pitch over it and the sea soon washed the Oakum out and the Water came under the Tower and down on the Berth Deck in Torents."

As the water continued to rise, the crew became convinced that the pounding was prying the upper and lower hulls apart, and that the widening cracks were admitting more and more water. Ericsson would afterward deny that that was possible, and put the full blame on the caulking around the base of the tower*.

Ericsson pointed out that "severe strain cannot take place in a structure nearly submerged." That's certainly true, but the point was that towing into a head sea, *Monitor* was not submerged, but rather rising out of the sea and crashing down again, exactly the kind of condition that might open up seams in her hull-to-deck joint.

The wind and seas continued to rise and the *Monitor* towed badly, yawing from side to side. The waves completely submerged the pilothouse and rolled over the turret as the ironclad drove her bows into the sea. This time, at least, only a little water came in through the blowers, not enough to do harm. The new, taller stacks had solved that problem

Around 9:00 P.M. Geer asked third assistant engineer Robinson Hands if he should prepare the large steam pump they had installed the previous May, and he was told to do so. It took him twenty minutes to ready the pump, and by then it was clear that the regular bilge pumps and the Worthington together were not keeping up with the water. Geer found Captain Bankhead on the top of the turret and reported the situation. Bankhead ordered the big pump fired up.

---

*Ericsson also laid a lot of the blame at Lieutenant Greene's feet. Greene was critical of a number of *Monitor*'s design characteristics, and though his criticism was well-founded, Ericsson went to great lengths to discredit anything that Greene said or did. Ericsson blamed Greene for letting the *Virginia* get away, and not following her and finishing her off, a very unfair charge.

The big pump was driven by two steam engines. After he helped in-stall it the previous spring, Geer had written to his wife, "It will pump 2500 Gallons pr minnit. It will keep us afloat if the hole was large enough for me to crall through." Now they would test that supposition.

For an hour or so the pump did keep up, flinging back overboard the huge quantities of water gushing in from various places, seen and un-seen. But it was only keeping up; it was not bringing the level of the wa-ter down. Meanwhile the storm's ferocity increased, and with it the influx of water.

Then, around 11:00 P.M., something let go. The water began to rise very fast, too fast for the pumps. Word was sent up to the officers hud-dled behind the iron bulwark that had been installed around the perimeter of the turret. The water was over the engine room floor and rising. "It was the death knell of the *Monitor*," Keeler recalled.

Just then the 11-inch hawser parted. There was no way to make it fast again. Nor did the *Monitor*'s crew have any way off the ship—their boats had been put aboard the *Rhode Island* for the trip south. Despite his great reluctance to admit defeat, Bankhead ordered that a signal for assistance be sent to *Rhode Island*.

After a brief consultation, the officers decided that it would be bet-ter to cut the remaining hawser away. Lieutenant Stodder and three seamen volunteered to go. Taking a hatchet, Stodder pulled himself along the deck, clinging to the lifelines that circled the perimeter of the deck. Again and again the heavy seas broke over the deck, submerging Stodder and the three men with him. Finally Stodder reached the bow, and with a few blows from the hatchet, the 15-inch-thick rope parted. Incredibly, Stodder managed to work his way back to the safety of the turret. All three men who had gone with him were swept away.

The men standing on the turret could see that *Rhode Island* was making reply to their signal, but they could not make out the meaning of the reply, and they could not tell if boats were being sent. Again and again the *Monitor*'s signal was repeated, and the men shouted against the wind, "Send your boats immediately, we are sinking!"

Bankhead ordered the anchor let go in hope that the drag would keep the *Monitor* bow to the wind, which it did. Keeler recalled that awful time in a letter to his wife: "Words cannot depict the agony of those moments as our little company gathered on the top of the turret, stood with a mass of sinking iron beneath them, gazing through the dim light . . . ," as they waited for *Rhode Island*'s boats to appear.

As they waited, the men were organized into a bucket brigade. This was more to give them something to do and stave off panic than in any hope that buckets could free the ship of water when a 2,500-gallon-per-minute pump could not. Down in the engine room, Geer found that the water was up to his knees and fast approaching the fireboxes. Soon it rose up over the engines driving the pump, and the pump came to a stop. "She was so full of Water, and roled and pitched so bad I was fearfull she would role under and forget to come up again," he wrote.

The *Rhode Island* had in fact understood *Monitor*'s distress signal. She stopped her engines immediately and launched three boats, though in those treacherous conditions the rescuers were in as much danger as those to be rescued. The boats clawed their way through the huge seas, disappearing between the high waves, then rising fast as the seas swept under them. Finally they approached near enough that they were seen from the *Monitor*'s turret. The *Monitor*'s men felt the first tinge of hope they had felt in some time.

As dangerous as it was for the boats to be in the open ocean, it was much worse coming alongside the waterlogged *Monitor*. With the ironclad and the boats rising and falling fast and often out of phase with one another, there was a good chance of the boats being crushed. As the first boat came alongside, a wave lifted it and dashed it down on *Monitor*'s deck, knocking a hole in the side. The next wave washed it off again. The boat crew stopped the leak as best they could and prepared to take on the *Monitor*'s men.

Lieutenant Greene and Commander Bankhead made their way down from the turret and onto the deck to assist the others in getting into the boats. There, with the huge seas breaking over them, clinging to the lifelines, they called for the members of the crew to follow. Not surprisingly, many were reluctant.

When Keeler saw the first boat appear out of the dark and the flying spray, it occurred to him that he might be able to save his account books. With an extraordinary and perhaps not so well-thought-out devotion to duty, Keeler climbed back into the turret, feeling his way around the guns, then, missing his step, fell through the hatch from the turret to the berth deck below.

There was one lantern still burning, casting the dimmest of light around the dank interior. The air below, which was generally humid, was now thicker than ever from the steam and gas of the extinguished fires in the engine room. Keeler pushed his way through the knee-high water and the semidarkness, forward toward the wardroom.

Following the *Monitor's* refit in Washington, the wardroom deck was lower than the berth deck, and when Keeler climbed down he found the water up to his waist and crashing side to side with the roll of the ship. Perhaps Keeler was not seaman enough to realize that that put the ship in imminent danger of capsize. Whether he knew it or not, his going below was an act of incredible courage.

The wardroom was pitch black, "a darkness that could be felt," but Keeler had hardly been off the ship in eleven months and could get around without seeing. In the "hot, stifling, murky atmosphere" he groped his way through the narrow passage to his stateroom. He gathered up the books and papers that he thought needed saving, but realized that they were almost too bulky to carry. There was no way he would get across the wildly pitching deck and into the boat carrying them, so he left them behind.

He managed to retrieve his watch, which was hanging from a nail, and then tried to open the safe in his cabin in which he kept the ship's store of "greenbacks." The safe was completely underwater, and Keeler was not able to get it open. Keeler gave that up, too, and headed back, with only his watch to show for his daring.

In the wardroom, the tables and chairs, floating in the chest-high water, surged across the space, threatening to knock Keeler senseless. He pushed on through, now very aware of the fact that the ship might at any moment take "the final plunge." Across the berth deck, up the ladder to the turret, feeling his way through the turret, over the tangle of gun tackles and the sponges and rammers and shot that had broken loose, Keeler finally emerged into the wild night on top of the turret. He arrived just in time to see the *Rhode Island*, out of control, come crashing into the *Monitor*.

Rhode Island

Onboard the *Rhode Island* they did not know that the *Monitor*'s crew had cut away the towing hawser. Maneuvering in the wild seas, *Rhode Island* had backed her engines and fouled the loose hawser in her port paddle wheel, temporarily disabling her. Being upwind of *Monitor*, *Rhode Island* began setting down fast on the ironclad.

Suddenly, out of the night, the men in *Monitor*'s turret saw their would-be rescuer coming down on them, stern first. A terrible collision seemed imminent; "it seemed certain that our iron edge would go through the big steamer's side and thus send us all to the bottom," Stodder recalled.

One of the rescue boats was alongside and the *Rhode Island* rode up against it, grinding and smashing it against *Monitor*'s side. The boat crew leapt onto *Monitor*'s treacherous deck to avoid being crushed. Taking advantage of the accident, men onboard the *Rhode Island* tossed lines over to *Monitor*'s deck, but none of *Monitor*'s men dared climb down for the turret onto the seaswept deck to grab for them.

The *Monitor*'s stern lay under the overhanging quarter of *Rhode Island*, and it needed only the right wave set to smash them together. Then, just as it looked as if the rescue ship would be the death of them all, the seas swept her past. She spun off downwind, driven much faster than the low and water-filled *Monitor*. Even the boat had escaped serious injury, with only the starboard gunnel crushed. Soon the *Rhode Island*'s crew had the paddle wheel free of the hawser and the engines turning again.

While Keeler was below, the first of the *Monitor*'s crew had begun making their way toward the boats, with Greene and Bankhead urging the reluctant men to leave the momentary safety of the turret for the plunging deck. In the dark night, with the sea surging around them, one by one they climbed down the steel ladder and staggered across the flat, open space.

But that was not the most dangerous moment. Worse than the open deck was the jump from deck to boat. The wild gyrations of the boat, rising and falling beside the sinking ironclad, made it nearly impossible to gauge the moment to jump. "[T]he boat sometimes held her place by the *Monitor*'s side, to be dashed hopelessly out of reach an instant later," Stodder wrote. There was nothing the men could do but jump and hope the boat would be there when they came down.

Often it was not. In a few cases, the men who landed in the water were hauled back into the boat. In many others, they disappeared into the night and the sea.

An old quartermaster made his way across the deck with a bundle under his arm, which Greene took to be his clothing. Greene shouted to the man that there was no time to save personal effects, but the quartermaster staggered on and threw the bundle, which turned out to be a ship's boy, into the rescue boat before jumping in himself.

George Geer was one of the last to get out of the ship. He climbed up onto the turret to find the rescue boats taking men away. Watching the men struggling across the open deck, he was not enthusiastic about making the attempt, but understood he might as well "be drowned trying to reach the Boats as to go down in the *Monitor*."

Geer made it down onto the main deck just as a big wave swept the ship fore and aft, knocking Geer down and tumbling him along. He managed to grab the lifeline, while the man beside him was swept overboard and lost. Once the water fell away, Geer pulled himself along until he managed to climb into the boat, coming "as close as I care to risking my life."

The *Monitor* was settling fast, her deck now level with the surface, and it says much about her stability—so doubted in the design phase—that she did not roll over with so much water in her. As the men were struggling for the boats, Bankhead ordered Stodder to go below and make certain no one was left behind.

In the wardroom, Stodder found Third Assistant Engineer Samuel A. Lewis in his bunk, too seasick to move. Lewis, described by Keeler as "a mere boy," had only joined the ship in November. Now Stodder told him that the ship was sinking and he had better get topside, but Lewis refused, too frightened and miserable to care. "[N]ot being strong enough to carry him," Stodder wrote, "I had to leave him to his fate." Lewis went down with the *Monitor*.

Captain Bankhead called to Paymaster Keeler to lead the reluctant men to the boats. Keeler shed his coat and some other clothing in case he had to swim. The turret's ladder had washed away, so he took hold of a rope hanging from an awning stanchion and slid from the turret to the deck.

As soon as his feet hit the wet iron, another wave rolled over the ship, tearing the rope from Keeler's hands, knocking him to the deck and tumbling him along. The sea swept Keeler clean overboard, and when he kicked his way to the surface he found he was 30 feet away from the ship. Then, incredibly, the backset of the wave whirled him back aboard the *Monitor*. He grabbed the lifeline and hung on, pulling himself along the deck. He reached the boat, and the men already aboard hauled him in.

Francis Butts had volunteered for the *Monitor* in November, while she was in Washington. ("I thought of what I had been taught in the service," he wrote about the *Monitor*'s sinking, "that a man always gets into trouble if he volunteers.") Now he found himself alone in the turret, the end of the bucket brigade from down below. His only companion was a black cat, which sat on one of the guns howling in fear. When he could stand it no longer, Butts pulled the tompion from the muzzle of the gun, stuffed the cat into the barrel and replaced the tompion, but alas he could still hear it.*

Finally there was no one on top of the turret to take the bucket from him, and Butts called for the men below to come up. Butts then climbed up onto the turret himself and slid down to the deck on a loose rope. As soon as he hit the deck he was washed away. Flailing around in the water, he managed to grab one of the smokestack braces, which he climbed hand over hand to get his head above water.

Then the water fell away and Butts found himself hanging from the cable. He dropped to the deck and grabbed the lifeline just as another wave swept the vessel, lifting Butts's feet off the deck. Slowly he worked himself along the lifeline and just managed to fling himself into the boat.

Finally there was only Commander Bankhead, Lieutenant Greene,

---

*Many historians doubt Butts's story of the cat, and certainly much of his account smacks of fabrication. Archaeological evidence does not support the story. As the conservator of the *Monitor*'s turret told this author, "I've had my hand down that barrel, and there ain't no cat."

and a handful of men still on the turret who, frozen with fear, refused to leave. Bankhead and Greene climbed into the last boat, and Bankhead spent some minutes begging the men in the turret to come down and attempt to reach them. But it was no use. They would not budge. Finally, the rescue boat shoved off, leaving the men behind. One hundred and forty years later, the skeletal remains of two of them would be found in the excavated turret.

The rescue boats, overloaded and leaking from having smashed against *Monitor*, had a pull of half a mile or more through the wild seas to reach *Rhode Island*. Incredibly, they all made it.

Once alongside, the men were once more faced with the peril of a small boat next to a ship in big seas, and the great danger of being swamped or crushed against *Rhode Island*'s side. Butts recalled, "We would rise upon the sea until we could touch her [*Rhode Island's*] rail; then in an instant, by a very rapid descent, we could touch her keel."

Ropes were tossed over *Rhode Island*'s side, but few of the men had the strength to pull themselves up onboard. Some just held on tight and were hauled aboard. As the boat rose up next to the *Rhode Island*, Butts grabbed one of the ropes dangling over the side and Ensign Norman Atwater grabbed another. Then the boat fell away from under them and was swept aft, leaving them dangling in the air. They hung there, sometimes 30 feet above the water as the ship rolled away. Finally Atwater could hold on no longer. With a cry of "O God!" his grip failed and he plunged into the sea and was never seen again.

Butts knew he was next. He could not hold onto the thin rope much longer. He called for help, but over the gale winds he was not heard. Finally, hope gone, he said a silent farewell to his family and friends. And then, incredibly, the boat rolled up under him like the hand of God and he fell into the bottom. Lying there, he heard an old sailor exclaim, "Where in————did he come from?"

Bankhead shouted for the men on *Rhode Island* to lower ropes with loops tied in the ends. Those men who could not climb, such as the *Monitor*'s new surgeon, Dr. Grenville Weeks, whose hand was crushed between the boat and the side of the *Rhode Island*, put their arms through the loops and were hauled onboard. Three of Weeks's fingers had to be amputated, a hard injury for a surgeon.

Finally, all the men who could be rescued had been. When they mustered on the *Rhode Island*'s deck, it was discovered that four officers and twelve crewmen were lost, swept away attempting to get to the boats, or too sick or frightened to even try.

It was then around one o'clock in the morning of December 31, 1862. The *Monitor*'s crew stood on the deck of the *Rhode Island* and watched the red and white lights hanging from the pennant staff on *Monitor*'s turret. Then the lights were gone, and the men knew that their beloved *Monitor* was as well. "The *Monitor* is no more," Keeler wrote to his wife. "What the fire of the enemy failed to do, the elements have accomplished."

Return to Hampton Roads

After the survivors of the *Monitor* had been deposited onboard the *Rhode Island*, one of the rescue boats, under the command of Acting Master's Mate D. Rodney Browne, volunteered to make one more run to the *Monitor* to try and retrieve those last few men. The boat was still making its way to the ironclad when the lights disappeared. Browne pressed on, hoping to find men in the water, but he could find no one.

By the time he was done looking, the *Rhode Island* was also out of sight and they were not able to find her. The last rescue boat was on its own in that wild sea.

When the boat failed to return, the *Rhode Island* began searching. They spent that night and the following day looking, but to no avail. Finally, thinking the boat lost, they gave up and steamed for Wilmington.

Incredibly, Browne and his crew survived the night intact. Browne made a sea-anchor of the boat's mast and trailed it over the bow, keeping the boat bow to the wind and seas. In the morning, they spotted a schooner standing toward them. Browne tied a black silk handkerchief to an oar and tried to signal, but, Hatteras having been a pirate base for hundreds of years, the schooner apparently took them for such and stood off. Finally they were picked up by the less timid *A. Colby* out of Bucksport, Maine, and brought into Beaufort, North Carolina, then in Union hands.

The *Monitor*'s men remained onboard *Rhode Island*, traveling to

Beaufort as well and finally back to Hampton Roads. During the voyage, the enlisted men composed a letter requesting a discharge from the navy. Upon arriving in Hampton Roads, the petition was forwarded to Washington, where it was later rejected. The U.S. navy was not about to dismiss experienced seamen just because they had suffered a shipwreck.

The *Monitor*'s officers were soon paid off and given leave, but the enlisted men were confined to the store ship *Brandywine*, where they had to "sleep on the Deck with no bedding or blankets" and make do with what little clothing they had come away from the *Monitor* with. It was shameful treatment, and the men endured that limbo for weeks. Finally they were paid one-fifth of what they were owed and allowed two weeks' leave.

Eventually the officers and men were reassigned and spread throughout the fleet. But they would all remain, for the rest of their lives, "the *Monitor* Boys."

The Old Jack Tar Feeling

"I share the old Jack Tar feeling that a sailor can do anything; and that a man is not good for much who is not a thorough seaman."

—GUSTAVUS VASA FOX

The loss of both ironclads, each less than a year after she was launched, was a moment of great sadness in the North and South.

Confederate naval-officer-turned-historian J. Thomas Sharf called *Virginia*'s destruction "the most distressing occurrence of the war up to that time," an impressive feat considering such Confederate losses as Shiloh and New Orleans. The Southern public had harbored wildly unrealistic expectations for their ironclad, and when she was destroyed, "public indignation knew no bounds."

Virginia's loss might have been more strongly felt than *Monitor*'s. For the Confederacy, *Virginia* represented a significant proportion of their ironclad fleet. Even if she was confined to Hampton Roads, she was a powerful weapon, her loss a blow to Southern strategy. That loss was felt particularly in Richmond. With Johnson's army trying to hold back the Union tide on the Peninsula, the *Virginia* was the only thing blocking the water route to the capital. According to one Southern historian, writing soon after the war, the destruction of *Virginia* "created a public grief, so wild and bitter, that at one time it was feared the building, in which were collected the departments of the government, might be stormed by a mob."

Her success, however, was an inspiration, and for the rest of the war, domestic Confederate naval shipbuilding focused primarily on ironclads. All Confederate ironclads followed roughly along the *Virginia* model, in part because they were mostly designed by John Luke Porter. They all featured a sloping iron casemate housing guns, both rifles and smoothbores, in broadside and on pivots. Some of *Virginia*'s traits were not copied. No other ironclad featured submerged ends (the innovation for which Brooke claimed credit) or casemates whose forward and after ends were rounded.

In 1865, the Confederate navy department authorized construction of a turreted ironclad in Columbus, Georgia, but the city fell to the enemy before construction could begin.

All told, the Confederacy laid down around fifty ironclads, of which twenty-two were completed and put into commission. Some, such as *Arkansas* and *Tallahassee,* did great service for the Confederacy. Most did not get a chance. But their psychological value was immense. Just as the Northern military was all but paralyzed with fear of *Virginia*, so "ram fever" would effect the U.S. navy wherever there was a Confederate ironclad in operation.

The Union's Loss

The *Monitor* "is a primary representative of a class identified with my administration of the Navy," Gideon Welles wrote nostalgically upon hearing of her sinking. As successful as she was, *Monitor* was, like *Virginia*, a prototype, and a flawed one at that.

Welles was slower than Mallory to take up the cause of ironclads, but he quickly saw their potential and embraced them. In December 1861, while *Monitor* was still under construction, Welles recommended to Congress that twenty more ironclad vessels be built, beginning immediately. Delays in Congress slowed up the authorization, but when Welles personally pushed for funding, Congress, in February 1862, voted ten million dollars for ironclad construction.

Though the Ericsson concept was still untried, the navy had already begun to see its benefit. In their published advertisement calling for iron-clad designs, the navy specified, without actually mentioning turrets, that the "guns of the vessels for harbor and coast defense are to train to all points of the compass without change in the vessel's position."

After *Virginia* made her deadly coming out, Northerners wanted to know where their ironclads were. New York diarist George Templeton Strong wrote, following the battle on March 8, "General dismay. What next? Why should not this marine demon [*Virginia*] breach the walls of Fortress Monroe, raise the blockade, and destroy New York and Boston? The nonfeasance of the Navy Department and of Congress in leaving us unprotected by ships of the same class, after ample time and abundant warning, is denounced by everyone."

The great Battle of Hampton Roads took place not long after Congress authorized the construction of more vessels, and the question was settled as to what sort of ironclads would be built. In fact, the public was so taken with the little *Monitor*'s success that Welles was "berated and abused" for not having "more vessels of the *Monitor* class under contract." From being ridiculed for investing in an absurdity such as Ericsson's Folly, Welles was now taken to task for not having built more of them!

Advertisements for building proposals that appeared in April called for "turrets on the plan of the Ericsson Turret." The navy also wanted more of the "Simon-pure article," entering into a contract with John Ericsson to design and build six single-turret monitors (the name of the original had by then become a generic for the type).

Gustavus Fox, who witnessed the *Monitor/Virginia* battle, decided to cut through the red tape to speed the process. He circumvented the Bureau of Construction and went over the heads of Chief Naval Constructor John Lenthall and Engineer-in-Chief Benjamin Isherwood, neither of whom were big Ericsson fans, and made arrangements with Ericsson directly.

The six new Ericsson ships, built and launched in 1862, were known as the *Passaic* class, and were larger and more heavily armed than the original *Monitor*, and had their pilothouses mounted on top of the turret. The *Passaic* herself, in fact, was steaming in consort with the *Monitor* when the latter foundered off Cape Hatteras. *Passaic* was nearly lost in the same storm.

Four others of the *Passaic* class were also built by other contractors to Ericsson's design. Ericsson was later contracted for two more monitors, double-turreted monsters called *Puritan* and *Dictator*.

Here again was a vivid demonstration of the North's overwhelming industrial capability. While Mallory was struggling to find engines and shafting and iron plate for the few disparate ironclads he had under construction, Welles was building dozens at a time. Besides the Ericsson monitors, 1862 saw the construction of twenty-eight other ironclad ships, a majority of them turreted. John Ericsson's genius had completely altered the way the navy thought of men-of-war.

Indeed, there was no one who emerged from the *Monitor* project with greater fame than John Ericsson. While the names of Worden, Greene, Stimers, and Newton became household words, their pictures everywhere, it was John Ericsson who came in for a lion's share of the adulation, both in the United States and internationally, and the stars aligned perfectly that that might happen.

Already discredited for the *Princeton* affair, Ericsson and his "Folly" had suffered from great abuse and ridicule while the *Monitor* was under construction. Most people expected very little of the Yankee Cheese Box; in fact, most did not think she would even float.

If *Monitor* had destroyed *Virginia* in dry dock, or even arrived in time to stop her destruction of March 8, and the Confederate ironclad's power had never been known, then perhaps little would have been made of Ericsson and his *Monitor*. Instead, *Monitor* arrived on the scene with cinematic timing.

The *Virginia*'s attack on March 8 sent the North into a panic. When the little *Monitor* charged in like the cavalry and stood up to the invincible *Virginia*, she was looked upon as the savior of the Union. She was the young David standing up to the dreaded Goliath (a comparison often made). And Ericsson went from being a goat to being a god.

He also went from struggling financially to being, if not fabulously rich, then certainly wealthy. Thanks to the fair and equitable deal orchestrated by Cornelius Bushnell, the profit Ericsson realized from the

Monitor was enough to set him up comfortably. Added to that was the money he made on building the eight monitor-type vessels for which he was personally responsible, as well as the royalties on those built by others. Within a few years, navies all over the world boasted monitors designed or inspired by Ericsson. From the summer of 1862 to the end of his life, John Ericsson would never again have to worry about money.

Needless to say, wealth did not slow a man of Ericsson's enormous energy. He continued to work on new developments in engine technology and naval warfare. In the 1870s he once again altered the nature of naval combat when he designed the *Destroyer*, the first ship to fire underwater torpedoes.

In February 1889, Ericsson received a shock when his dear friend and partner Cornelius Delameter passed away at the age of sixty-eight, eighteen years Ericsson's junior. Despite growing infirmity, Ericsson continued to work as long as he could stand at the drafting table. Finally, in March of that year, his health began to decline rapidly. Laid comfortably in his bed, the captain said, "I am resting. This rest is magnificent; more beautiful than words can tell." Thirty-nine minutes past midnight, at eighty-six years of age, John Ericsson breathed his last. The date was March 8.

Cornelius Bushnell also profited greatly from the ironclad business. Though *Galena* was a failure as an ironclad, Bushnell continued as Ericsson's partner in subsequent monitors, continuing to watch the business end as Ericsson, whom he now considered a close friend, concentrated on engineering.

Bushnell continued his railroad speculation, helping found the Union Pacific Railroad that linked the east and west coasts. He made a fortune on that deal, though he would lose much of it in less profitable speculation on other railroads, as well as iron and coal mines. His part in the creation of the *Monitor* was well recognized, and he carried that honor with him until the end.

Gideon Welles served as secretary of the navy throughout the war, one of only two cabinet members to do so. He continued to serve until the end of the Johnson administration in 1869. During his eight-year tenure he transformed the U.S. navy from the sleepy, inbred service he inherited to one of the most powerful naval forces in the world, second only to Great Britain's. He was sixty-six when he retired from politics and returned to Hartford, Connecticut, where he lived until his death in February 1878.

The men of the *Monitor*, or at least the officers, were lionized by the Northern press and the public. The entire crew received the Thanks of Congress. Worden was promoted to commander a few months after the battle and captain the following year. His next command was the new monitor *Montauk*, in which he fought with the South Atlantic Blockading Squadron. In 1863 Worden was sent to New York and spent the remainder of the war working on the development of ironclad warships.

During his fifty-year navy career, John Worden served as superintendent of the U.S. Naval Academy and commanded the European squadron with the rank of admiral. He retired in 1886 to New York, and died there eleven years later.

Worden never regained sight in one eye from the hit he took from *Virginia*'s gun, and shards of metal from the pilothouse remained imbedded in his skin for the rest of his life. Five decades after the fight, his skin was described as "mottled with blue . . . On the left side of his nose was a slight depression, in which could be felt a small piece of steel." John Worden literally took a part of the *Monitor* with him to the grave.

Samuel Dana Greene served as executive officer aboard the USS *Florida* and the *Iroquois* during the remainder of the war, receiving promotion to lieutenant commander in 1865 and then commander in 1871. He held various commands and taught mathematics and astronomy at the U.S. Naval Academy. In 1884, while serving as executive officer at the Portsmouth Naval Shipyard, Greene committed suicide. He was forty-five years old.

Immediately following the battle, Alban Stimers enjoyed a highpoint in his reputation and fame. The volunteer engineer who ran the turret mechanism was credited, along with Worden, with the *Monitor*'s success.

When Fox arranged for Ericsson to design and build the new monitors, he also arranged for Stimers to assist. Stimers set up offices next door to Ericsson and for a while the two men worked well together, collaborating on the *Passaic* class ships, and then the nine ship *Canonicus* class of monitor. It was almost inevitable that two headstrong engineers such as Ericsson and Stimers would eventually clash, and finally they did. A falling-out over the new class of ship ended their working relationship.

After serving as fleet engineer for Admiral Samuel Du Pont's failed attempt to take Charleston in April 1863, Stimers returned to New York City to continue work on the new *Casco* class, his first independent project, a monitor designed for river warfare.

Fox was under the impression that Stimers and Ericsson were still cooperating with one another, but in fact Stimers was on his own. When a young assistant miscalculated the displacement of the new ironclads, the mistake went unnoticed. When the first of the *Casco*s was launched, she floated a foot deeper than she was intended. There was no chance she would bear the weight of turret, guns, coal, and crew. It was a monumental blunder, rendering all twenty of the *Casco* class (the construction of which Stimers had seriously mismanaged) nearly useless.

That was the end of Stimers's career in ship design. He was given a series of increasingly humiliating shipboard assignments until at last he quit the navy in 1865. He joined his old *Monitor* shipmate Isaac Newton in civilian steam engineering in New York City. He died in 1876 of smallpox, largely forgotten.

Paymaster Keeler joined Samuel Greene on the blockade ship *Florida*, on which he served for the remainder of the war. A wound in the back from a shore battery earned him a ten-dollar-per-month pension after leaving the navy in 1865. A few years after the war, Keeler moved to Florida, where he was engaged in various enterprises, none of them terribly profitable. He did not discuss his naval background, and for some reason appropriated the rank of Major, by which he became known locally. He died of heart disease in 1886.

Confederate Veterans

In the South, the men of the *Virginia* were as famous and revered as their *Monitor* counterparts were in the North, receiving the thanks of the Confederate Congress. Foremost among them, of course, was Franklin Buchanan.

Buchanan spent two months bedridden from his wound. Miraculously, the Yankee bullet had missed the femoral artery, as well as the femur and pelvis. A hit to any of those would probably have killed the sixty-one-year-old captain. But the bullet still managed to do serious damage, and Buchanan never recovered entirely from the wound.

Buchanan spent his convalescence receiving the thanks and praise of friends and admirers from all over the Confederacy. Josiah Tattnall wrote that he hoped "Congress will make you an admiral and put you at the head of our navy." And they very nearly did.

In August 1862 Buchanan was promoted to full admiral, thanks to Mallory's policy of promoting by merit and not seniority (Buchanan protested that Tattnall should be made a full admiral first, due to seniority, but Mallory ignored the suggestion). In fact, Buchanan would be the only full admiral in the Confederate navy.

When at last Old Buck had recovered enough to resume active duty, he was assigned to the defense of Mobile Bay, where he took command of the new ironclad flagship *Tennessee*. On August 5, 1864, Buchanan led the Confederates in a pitched battle against Admiral Farragut's fleet (Buchanan had earlier written to Robert Minor "I should like to have had a crack at my old *friend* and *messmate* [Farragut]"). In the end, only *Tennessee* was left fighting, but she too was so shot up that she was forced to surrender. Buchanan, wounded even more severely than he had been aboard *Virginia*, was taken prisoner.

By the time Buchanan was exchanged, the war was nearly over, and he did not see any further active duty. With the end of the war, money became a problem for Buchanan, since the navy pension he had counted on was lost with his resignation from the U.S. navy. He supported himself and his wife through small-time farming until he was offered the position of president of the Agricultural College of Maryland, the forerunner of the University of Maryland.

Old Buck had not mellowed since the war. Writing of his duties to a friend and former Confederate naval officer, Buchanan said, "At first I had to practice a little *man* of *war* discipline." He finally resigned the post when the board of trustees would not uphold his decision to fire a professor whom he felt was unqualified. After another stint as a farmer, Buchanan accepted a position selling life insurance (an occupation to

which many former Confederate officers turned after insurance companies discovered the promotional value of their names). He returned
briefly to Mobile, this time as a salesman.

Leaving that distasteful work after a year and a half, Buchanan returned to his beloved Maryland. He died of pneumonia in May 1874.
He was buried with two iron cannonballs, souvenirs he had brought
back with him from the Mexican War, a happier time when he and Farragut, Tattnall, and Goldsborough were all fighting side by side.

As dismayed as all the Confederacy was with the destruction of *Virginia*, it was natural that much of the blame would fall on Josiah Tattnall. A court of enquiry found that "The destruction of the *Virginia* was,
in the opinion of the court, unnecessary at the time and place it was ef-
fected," and that Tattnall was to blame for not having at least taken her
as far upriver as Hog Island and used her for defense of the James.

Many officers were outraged at this finding, including Buchanan,
and not least of all Josiah Tattnall. Tattnall demanded and got a court-
martial, which saw the evidence differently. At the end of lengthy testimony the court gave Tattnall an honorable acquittal.

At sixty-seven, Tattnall was too old for sea service. He was assigned
to the shore defenses of Savannah, Georgia, and became a prisoner of
war when the city fell to the Union. He was not paroled until 1865. After the war he moved to Nova Scotia, returning, impoverished, to Savannah in 1870. He died there a year later.

Catesby ap Roger Jones went on to command the ironclad *Chatta-
hoochie*. In 1863 he was promoted to commander and put in charge of
the Confederate foundry and ordnance works in Selma, Alabama.
There he produced guns and shot for the ships of the Confederate
navy, including Buchanan's *Tennessee*. After the war he settled in
Selma and pursued various business interests. He was killed in 1877 in
a confrontation with a friend. He was fifty-six years old.

Like his Union counterpart, Stephen R. Mallory was one of only
two cabinet members to serve throughout the war. Unlike Welles, Mallory's navy was always starved for money, men, and matériel, but no one
could have done better with what he had. His policies of ironclad construction and commerce raiding delivered the biggest hurt possible to
an enemy of overwhelming strength.

When Richmond fell, Mallory abandoned the city with the rest of the government. He was captured in Georgia in May 1865 and imprisoned for ten months. Following his release, he returned to the practice of law, living in Pensacola, Florida, until his death in 1873 at the age of sixty.

And the Winner Is . . .

So who won the great battle? Most observers then and now must conclude that the actual fight between the ironclads was a draw. In a letter to Mallory, Buchanan wrote, "The *Monitor* is from the best information I can obtain from the officers of this ship [*Virginia*] her equal." John Taylor Wood concluded, "The battle was a drawn one, so far as the two vessels engaged were concerned."

Likewise, the Northern observers were not quick to dismiss *Virginia's* power. Gustavus Fox telegraphed McClellan, "The performance of the *Monitor* to-day against the *Merrimack* shows a slight superiority in favor of the *Monitor*," an opinion held by many and based no doubt on *Monitor's* superior maneuverability. But Fox acknowledged that *Virginia* had steamed away under her own power and was probably not much hurt. "She is an ugly customer, and it is too good luck to believe we are yet clear of her," he concluded.

In 1875, John Worden and the officers of the *Monitor* petitioned Congress for $200,000 in prize money for "damage to the Confederate ironclad *Merrimac*, March 9, 1862, and her subsequent destruction." It was an absurd claim. As many of *Virginia's* officers pointed out, the only real damage done to the *Virginia* had been done by *Cumberland* the day before, or by *Virginia* herself in ramming the *Cumberland*.

Absurd or not, it raised a howl of protest, and launched a thousand letters exploring every aspect of the battle and the reasons that *Monitor* most certainly did not "whip" *Virginia*. Worden's petition dragged on for years. In the end, the Senate approved it but the House did not, and prize money was never awarded.

Among historians there is also some disagreement as to who the winner was. Certainly, looking at both days of battle, the 8th and the 9th, *Virginia* was the clear winner, considering the punishment she doled out to the U.S. navy. But on the 9th, she was thwarted in her effort to destroy the *Minnesota*. Though *Monitor* did little damage to the Confederate ship, the fact remains that *Virginia* steamed off for dry dock, much battered, while *Monitor* remained at *Minnesota*'s side.

Some have argued that *Virginia* was the overall winner strategically, since her presence denied the Union Hampton Roads and the use of the James River. She was not formidable enough, however, to deny them the use of Fortress Monroe and Old Point Comfort. The *Monitor* and the rams were too much of a threat for her to wander into their neighborhood. The loss of the James River was an inconvenience for McClellan, but it did not stop him from attacking Richmond via the Peninsula.

Ultimately it was a stalemate. *Virginia* held her corner of Hampton Roads, *Monitor* hers. Neither cared to go into the other's territory. Each thought the other was afraid to fight.

But a stalemate such as that favored the Union, because the Union could go by another route, could move supplies up a river that the *Virginia* could not reach. The *Virginia* held Hampton Roads, but in the end, it did not matter.

The legacy of the *Monitor* and the *Virginia* does not rest on the question of the ships that won, but rather the ships that lost. And the ships that lost were the wooden walls, the fleets of the world, the end products of thousands of years of development. Like so much of human history, they were swept away with staggering speed by the rise of the industrial revolution. On March 8, 1862, the sailors and shipwrights of the world took a step back, and into the foreground came the engineers, ushering in the reign of iron.

Notes

Introduction

Campbell and Flanders, *Confederate Phoenix*; Cline, *Ironclad Ram Virginia*; Flanders, *Memoirs of Jack*; Wise, *End of an Era*.

Chapter One

Besse, *Virginia and Monitor*; J. Brooke, *Her Real Projector*; Buchanan, *Letterbook*; Campbell and Flanders, *Confederate Phoenix*; Cline, *Ironclad Ram*; Curtis, *History*; Eggleston, *Narrative*; Flanders, *Merrimac*; Flanders & Westfall, ed., *Memoirs*; Jones, *Services*; Newton, *Merrimack or Virginia*; *Official Records of the Navies*; Porter, *Plan and Construction*; Porter, *Virginia, Her Story*; Ramsay, *Narrative*; *San Francisco Chronicle*, "Story of Peterson"; Selfridge, *Memoirs*; Shippen, *Thirty Years*; Symonds, *Confederate Admiral*; Wise, *End of an Era*; Wood, *First Fight*; Worden et al., *Monitor and Merrimac*.

Chapter Two

Daly, *Aboard the* Monitor; Dunbrow, *Stimers*; Greene, *In the* Monitor's *Turret, Letter*; *Monitor* Log; Navy Department, *Civil War Naval Chronology*; *Official Records of the Navies*; Stodder, "Ten Months"; Welles, *Diary*; *Report of Captain Worden*, Worden Papers.

Chapter Three

Bassler, *Collected Works*; Denny, *Civil War Years*; Flanders, *Merrimac*;

McPherson, *Battle Cry*; Musicant, *Divided Waters*; Navy Department, *Civil War Naval Chronology*; *Official Records of the Navies*; 37th Congress, 2nd session, Senate Report No. 37, *Surrender and Destruction of Navy Yards, Etc.*; Welles, *Diary*.

Chapter Four

Bennet, *Steam Navy*; Denny, *Civil War Years*; Navy Department, *Civil War Naval Chronology*; *Official Records of the Navies*; 37th Congress, 2nd session, Senate Report No. 37, *Surrender and Destruction of Navy Yards, Etc.*; Selfridge, *Memoirs*; Welles, *Diary*.

Chapter Five

Denny, *Civil War Years*; *Official Records of the Navies*; Flanders, *Merrimac*; Navy Department, *Civil War Naval Chronology*; 37th Congress, 2nd session, Senate Report No. 37, *Surrender and Destruction of Navy Yards, Etc.*; Selfridge, *Memoirs*; Sharf, *Confederate Navy*; Welles, *Diary*.

Chapter Six

Denny, *Civil War Years*; McPherson, *Battle Cry*; Navy Department, *Civil War Naval Chronology*; *Official Records of the Navies*; 37th Congress, 2nd session, Senate Report No. 37, *Surrender and Destruction of Navy Yards, Etc.*; Selfridge, *Memoirs*; Sharf, *Confederate Navy*; Welles, *Diary*.

Chapter Seven

Durkin, *Mallory*; Luraghi, *Confederate Navy*; McPherson, *Battle Cry*; Musicant, *Divided Waters*; Navy Department, *Civil War Naval Chronology*; *Official Records of the Navies*; Sharf, *History*; Still, ed., *The Confederate Navy*; Wells, *Confederate Navy*.

Chapter Eight

Baxter, *Introduction of the Ironclad*; Bennett, *Steam Navy*; Brooke, *John M. Brooke*; Brooke, *Virginia or Merrimac*; Durkin, *Mallory*; Luraghi, *Confederate Navy*; *Official Records of the Navies*; Paine, *Warships*; Sharf, *Confederate States Navy*; Smith, *Iron and Heavy Guns*; Still, ed., *Confederate Navy*; Wise, *End of an Era*.

Chapter Nine

American National Biography; Basler, *Collected Works*; Bennett, *Steam Navy*; Bushnell, *Negotiations*; Bushnell, *Paper to Army Navy Club*; McCordock, *Yankee Cheese Box*; Senate Journal, 37th Congress, 1st session; XXXVIII; Welles, *Diary*; Welles, *"First Ironclad"*; Wells, *Story of Monitor*; West, *Gideon Welles*.

Chapter Ten

Baxter, *Introduction of the Ironclad*; Bennett, *Steam Navy*; J. Brooke, *John M. Brooke*; J. Brooke, *Plan and Construction*; Campbell and Flanders, *Confederate Phoenix*; Dew, *Ironmaker*; Durkin, *Mallory*; Flanders, *Merrimac*; Forrest, *Letterbook*; Luraghi, *Confederate Navy*; *Official Records of the Navies*; Porter, *Plan and Construction*; Porter, Virginia, *Story of Her Construction*; Sharf, *Confederate States Navy*; Smith, *Iron and Heavy Guns*; Still, ed., *Confederate Navy*; Wells, *Confederate Navy*.

Chapter Eleven

Baxter, *Introduction of the Ironclad*; Bennett, *Steam Navy*; Bushnell, *Negotiations*; Bushnell, *Paper to Army Navy Club*; Church, *Ericsson*; McCordock, *Yankee Cheese Box*; *Official Records of the Navies*; Welles, *Diary*; Welles, *"First Ironclad"*; Wells, *Story of Monitor*; West, *Gideon Welles*.

Chapter Twelve

Baxter, *Introduction of the Ironclad*; Besse, *Virginia and Monitor*; G. Brooke, *Brooke*; J. Brooke, *Papers*; J. Brooke, *Plan and Construction*; Brooke, *The Real Projector*; Campbell and Flanders, *Confederate Phoenix*; Dew, *Ironmaker*; Flanders, *Merrimac*; National Archives; *Official Records of the Navies*; Porter, *Plan and Construction*; Porter, *Virginia, Her Story*; Smith, *Iron and Heavy Guns*; Still, *Confederate Shipbuilding*; Wise, *End of an Era*.

Chapter Thirteen

Baxter, *Introduction of the Ironclad*; Bennett, *Steam Navy*; Church, *Ericsson*; Fox, *Correspondence*; White, *Yankee from Sweden*.

Chapter Fourteen

Baxter, *Ironclad Warship*; Bennett, *Steam Navy*; Church, *Ericsson*; White, *Yankee from Sweden*.

Chapter Fifteen

Baxter, *Introduction of the Ironclad*; Bennett, *Steam Navy*; Church, *Ericsson*; Welles, *Diary*; Welles, *"First Ironclad"*; Wells, *Story of Monitor*; West, *Gideon Welles*; White, *Yankee from Sweden*.

Chapter Sixteen

Baxter, *Introduction of the Ironclad*; Besse, *Virginia and Monitor*; G. Brooke, *Brooke*; J. Brooke, *Papers*; J. Brooke, *Plan and Construction*; J. Brooke, *The Real Projector*; Campbell and Flanders, *Confederate Phoenix*; Dew, *Ironmaker*;

Flanders, *Merrimac*; Forrest, *Letterbook*; McPherson, *Battle Cry*; *Official Records of the Navies*; Porter, *Plan and Construction*; Porter, *Virginia, Her Story*; Ramsay, *Narrative*; Smith, *Iron and Heavy Guns*; Still, *Confederate Shipbuilding*; Still, ed., *Confederate Navy*; Wise, *End of an Era*; Woodward, ed., *Chesnut's Civil War*; Worden et al., *Monitor* and *Merrimac*.

Chapter Seventeen

Boynton, *History of the Navy*; Bushnell, *Negotiations*; Bushnell, *Paper to Army and Navy Club*; Church, *Ericsson*; Ericsson, *Building the Monitor*; Fox, *Correspondence*; Newton, *Shot-Proof Vessels*; Luraghi, *Confederate Navy*; National Archives, *Contract for Building the Monitor*; Still, *Monitor Builders*; Welles, *Diary*; White, *Yankee from Sweden*.

Chapter Eighteen

Bushnell, *Paper to the Army and Navy Club*; Church, *Ericsson*; National Archives, *Monitor Contract*; Porter, *Incidents and Anecdotes*; Scientific American, *"Ericsson Battery"*; Still, *Monitor Builders*; Welles, *First Ironclad Monitor*; Welles, *Diary*; White, *Yankee from Sweden*.

Chapter Nineteen

Baxter, *Introduction of the Ironclad*; Besse, *Virginia and Monitor*; G. Brooke, *Brooke*; J. Brooke, *Papers*; J. Brooke, *Plan and Construction*; J. Brooke, *The Real Projector*; Campbell and Flanders, *Confederate Phoenix*; Dew, *Ironmaker*; Flanders, *Merrimac, Diary of Jack*; Forrest, *Letterbook*; C. Jones, *Services*; *Official Records of the Navies*; Porter, *Plan and Construction*; Porter, *Virginia, Her Story*; Ramsay, *Narrative*; Smith, *Iron and Heavy Guns*; Still, *Confederate Shipbuilding*; Still, ed., *Confederate Navy*; Worden et al., *Monitor and Merrimac*.

Chapter Twenty

Besse, *Virginia* and *Monitor*; Church, *Ericsson*; Daly, *Aboard the Monitor*; Ericsson, *Building of the "Monitor"*; Forrest, *Letterbook*; Konstam, *Union Monitor*; Marvel, *Monitor Chronicles*; McCordock, *The Yankee Cheese Box*; *Monitor Contract*; Newton, *Shot-proof Vessels*; *ORN*, Series II, Volume 2; *Scientific American, The Ericsson Battery*; Still, Monitor *Builders*; Worden Papers.

Chapter Twenty-One

Besse, *Virginia* and *Monitor*; G. Brooke, *Brooke*; J. Brooke, *Papers*; J. Brooke, *Plan and Construction*; J. Brooke, *The Real Projector*; *Official Records of the Navies*; Campbell and Flanders, *Confederate Phoenix*; Dew, *Ironmaker*; Eggleston, *Narrative*; Flanders, *Merrimac*; Forrest, *Letterbook*; Porter, *Plan and Construction*; Porter, *Virginia, Her Story*; Ramsay, *Narrative*; Sharf, *Confederate*

Navy; Still, *Confederate Shipbuilding*; Wood, *First Fight*; Worden et al., *Monitor and Merrimac.*

Chapter Twenty-Two
Bennett, *Steam Navy*; Besse, Virginia *and* Monitor; Church, *Ericsson*; Ericsson, *Building of Monitor*; Gosnell, *Guns*; "Iron-clad Vessels,"; McCordock, *Yankee Cheese Box*; Monitor *Contract*; Musicant, *Divided Waters*; National Archives; Navy Department, *Civil War Naval Chronology*; *Official Records*; Porter, *Incidents and Anecdotes*; Stimers, Monitor *and Alban Stimers*; Welles, *First Ironclad*, Monitor.

Chapter Twenty-Three
Besse, *Virginia and Monitor*; G. Brooke, *Brooke*; J. Brooke, *Papers*; J. Brooke, *Plan and Construction*; J. Brooke, *The Real Projector*; Buchanan, *Letterbook*; Buchanan, Letter to Du Pont, Campbell and Flanders, *Confederate Phoenix*; Cline, *Ironclad Ram*; Dew, *Ironmaker*; Eggleston, *Narrative*; Flanders, *Merrimac*; Forrest, *Letterbook*; Jones, *Services*; Navy Department, *Civil War Naval Chronology*; *Official Records of the Navies*; Porter, *Plan and Construction*; Porter; *Virginia, Her Story*; Ramsay, *Narrative*; Still, *Confederate Shipbuilding*; Still, ed., *Confederate Navy*; Wise, *End of an Era*; Wood, *First Fight*; Worden et al., *Monitor and Merrimac.*

Chapter Twenty-Four
Boynton, *History of the Navy*; Navy Department, *Civil War Naval Chronology*; Daly, *Aboard the* Monitor; Durbrow, *Stimers*; McPherson, *Battle Cry*; Musicant, *Divided Waters*; *Official Records of the Navies*; Ould, *Exchange of Prisoners*; Smith, *Iron and Heavy Guns*; Welles, *Diary*; Worden Papers.

Chapter Twenty-Five
G. Brooke, *Brooke*; J. Brooke, *The Real Projector*; Cline, *Ironclad Ram*; Campbell and Flanders, *Confederate Phoenix*; Eggleston, *Narrative*; Flanders, *Merrimac*; Jones, *Services*; Navy Department, *Civil War Naval Chronology*; Norris, *Story of Virginia*; *Official Records of the Navies*; Porter, *Plan and Construction*; Porter, *Virginia, Her Story*; Ramsay, *Narrative*; Still, *Confederate Shipbuilding*; Still, *Iron Afloat*; Still, ed., *Confederate Navy*; Symonds, *Confederate Admiral*; Wise, *End of an Era*; Wood, *First Fight*; Worden et al., *Monitor and Merrimac.*

Chapter Twenty-Six
Daly, *Aboard the* Monitor; Durbrow, *Stimers*; Greene, *In the* Monitor's *Turret*; Marvel, ed., *Monitor Chronicles*; McCord, "*Ericsson and his Monitors*"; Moni-

tor Log, National Archives; Newton, *Shot-Proof Vessels; Official Records of the Navies;* Smith, *Iron and Heavy Guns.*

Chapter Twenty-Seven

Bennett, *Steam Navy;* Besse, *Virginia and Monitor;* J. Brooke, *The Real Projector;* Buchanan, *Letterbook;* Cline, *Ironclad Ram;* Curtis, *History;* Eggleston, *Narrative;* Flanders, ed., *Memoirs;* Jones, *Services;* Newton, *Merrimack or Virginia; Official Records of the Navies;* Parker, *Recollections;* Ramsay, *Narrative;* Selfridge, *Memoirs;* Shippen, *Thirty Years;* Symonds, *Confederate Admiral;* Wise, *End of an Era;* Wood, *First Fight;* Worden et al., *Monitor and Merrimac.*

Chapter Twenty-Eight

Bennett, *Steam Navy;* Besse, *Virginia and Monitor;* J. Brooke, *The Real Projector;* Buchanan, *Letterbook;* Cline, *Ironclad Ram;* Curtis, *History;* Eggleston, *Narrative;* Flanders, ed., *Memoirs;* Jones, *Services;* Newton, *Merrimack or Virginia; Official Records of the Navies;* Parker, *Recollections;* Ramsay, *Narrative;* Selfridge, *Memoirs;* Shippen, *Thirty Years;* Symonds, *Confederate Admiral;* Wise, *End of an Era;* Wood, *First Fight;* Worden et al., *Monitor and Merrimac.*

Chapter Twenty-Nine

Colston, "*Watching the* Merrimac"; Curtis, *Congress and the Merrimac;* Daly, *Aboard the* Monitor; Durbrow, *Stimers;* Eggleston, *Narrative;* Greene, *In the* Monitor's *Turret, Letter;* Jones, *Services; Monitor* Log; *Official Records of the Navies;* Phillips, *Notes;* Shippen, *Thirty Years;* Stodder, "*Ten Months*"; Wise, *End of an Era;* Wood, *First Fight;* Worden, et al., Monitor *and* Merrimac; Worden Papers.

Chapter Thirty

Colston, "*Watching the* Merrimac"; *Monaghan Papers;* Niven, *Welles; Official Records of the Navies;* Welles, *Diary;* Welles, *First Ironclad;* Wise, *End of an Era;* Woodward, ed., *Chesnut's Civil War.*

Chapter Thirty-One

Colston, "*Watching the* Merrimac"; Curtis, *Congress and the Merrimac;* Daly, *Aboard the* Monitor; Durbrow, *Stimers;* Eggleston, *Narrative;* Greene, *In the* Monitor's *Turret, Letter;* Jones, *Services;* Littlepage, Merrimac *vs.* Monitor; *Monitor* Log; *Official Records of the Navies;* Phillips, *Notes;* Stodder, "*Ten Months*"; Wise, *End of an Era;* Wood, *First Fight;* Worden et al., *Monitor and Merrimac;* Worden Papers.

Chapter Thirty-Two

Colston, "*Watching the* Merrimac"; Curtis, *Congress and the Merrimac;* Daly, *Aboard the* Monitor; Davis, *Duel Between the Ironclads;* Durbrow, *Stimers;* Eggleston, *Narrative;* Ericsson, *Building of the* Monitor; Greene, *In the* Monitor's *Turret, Letter;* Jones, *Services;* Konstam, *Union Monitor;* Littlepage, *Merrimac vs. Monitor; Monitor* Log; Milligan, "*Engineer Aboard* Monitor"; *Official Records of the Navies;* Ramsay, *Narrative;* Stodder, *Ten Months;* Swinton, *Monitor and Merrimac;* Wise, *End of an Era;* Wood, *First Fight;* Worden et al., *Monitor and Merrimac; Report of Captain Worden,* Worden Papers.

Chapter Thirty-Three

Colston, "*Watching the* Merrimac"; Curtis, *Congress and the Merrimac;* Daly, *Aboard the* Monitor; Davis, *Duel Between the Ironclads;* Durbrow, *Stimers;* Eggleston, *Narrative;* Ericsson, *Building of the* Monitor; Greene, *In the* Monitor's *Turret, Letter;* Jones, *Services;* Konstam, *Union Monitor;* Littlepage, *Merrimac vs. Monitor;* Milligan, "*Engineer Aboard Monitor*"; *Monitor* Log; *Official Records of the Navies;* Ramsay, *Narrative;* Stodder, *Ten Months;* Swinton, *Monitor and Merrimac;* Wise, *End of an Era;* Wood, *First Fight;* Worden et al., *Monitor and Merrimac; Report of Captain Worden,* Worden Papers.

Chapter Thirty-Four

Curtis, *Congress and the Merrimac;* Daly, *Aboard the* Monitor; Davis, *Duel Between the Ironclads;* Durbrow, *Stimers;* Eggleston, *Narrative;* Fox, *Confidential Correspondence;* Greene, *In the* Monitor's *Turret, Letter;* Jones, *Services;* Konstam, *Union Monitor;* Marvel, ed., *Monitor Chronicles;* Milligan, *Engineer Aboard Monitor; Monitor* Log; *Official Records of the Navies;* Ramsay, *Narrative;* Stodder, *Ten Months;* Swinton, *Monitor and Merrimac;* Welles Papers; Wise, *End of an Era;* Wood, *First Fight; Report of Captain Worden,* Worden Papers.

Chapter Thirty-Five

Basler, *Works of Lincoln;* Buchanan Letterbook; Daly, *Aboard the Monitor;* Davis, *Duel;* Foute, *Echoes;* Jones, *Diary;* Jones, *Services;* Littlepage, *Merrimac vs. Monitor;* Marvel, *Monitor Chronicles;* Newton, *Merrimac;* Newton Papers; *Official Records of the Navies;* Parker, *Recollections;* Porter, *Plan and Construction;* Sandburg, *Lincoln;* Sharf, *Confederate Navy;* Smith, *Iron;* Still, ed., *Confederate Navy;* Wood, *First Fight.*

Chapter Thirty-Six

Curtis, *History;* Davis, *Duel;* Denny, *Civil War Years;* Foute, *Echoes;* Jones, *Services;* Littlepage, *Merrimac vs. Monitor;* Newton, *Merrimac;* Newton Papers; *Official Records of the Navies*; Parker, *Recollections*; Sharf, *Confederate Navy*; Smith, *Iron*; Still, ed., *Confederate Navy*; Wood, *First Fight*; Worden et al., *Both Sides*.

Chapter Thirty-Seven
Butts, *Loss of Monitor*; Daly, *Aboard the Monitor*; Davis, *Duel*; Denny, *Civil War Years*; Ericsson, *Building the Monitor*; Greene, *In the Monitor's Turret*; Marvel, *Monitor Chronicles*; *Official Records of the Navies*; Smith, *Iron*; Stodder, *Ten Months*; Worden et al., *Both Sides*.

Epilogue
American National Biography; Bennett, *Steam Navy*; Church, *Ericsson*; Daly, *Aboard the Monitor*; Davis, *Duel*; Konstam, *Union Monitor*; Luraghi, *Confederate Navy*; *Minor Papers*; National Archives; *Official Records of the Navies*; Sharf, *Confederate Navy*; Still, ed., *Confederate Navy*; Still, *Iron Afloat*; Strong, *Diary*; Symonds, *Confederate Admiral*; Wegner, *"Ericsson's High Priest"*; Welles, *Diary*.

Selected Bibliographies

Virginia

Manuscripts

Brooke, John M. *John M. Brooke Papers*. Southern Historical Collection, University of North Carolina, Chapel Hill.

Buchanan, Franklin. *Franklin Buchanan Letterbook*. Southern Historical Collection, University of North Carolina, Chapel Hill.

Buchanan, Franklin, Buchanan to Samuel F. Du Pont, May 20, 1861 Samuel Francis Dupont Collection, Hagley Museum and Library, Wilmington, Delaware

Forrest, French. *French Forrest Letterbook*. Southern Historical Collection, University of North Carolina, Chapel Hill.

Minor Family Papers. Virginia Historical Society, Richmond, Virginia.

Porter, John Luke. C.S.S. Virginia (Merrimack) *Story of Her Construction, Battles, Etc.* (copy). Mariners' Museum Collection, Newport News, Virginia.

———. *Merrimac Gundeck*. Original drawing in the Mariners' Museum Collection, Newport News, Virginia.

Ramsey, H. Ashton. *Narrative of H. Ashton Ramsey, Chief Engineer, Confederate States Steamer* Merrimack, *During Her Engaguement in Hampton Roads—1862*. Record Group 45, National Archives, Washington, DC.

Published Primary Sources

Brooke, John Mercer. *The Plan and Construction of the Merrimac, Battles and Leaders of the Civil War*, vol. 1, R. U. Johnson and C. C. Clough Buel, ed. New York: The Century Company, 1887.

————. *The* Virginia *or* Merrimac: *Her Real Projector*. Richmond, Virginia: *Southern Historical Society Papers* XX, (1891).

Cline, William R. *The Ironclad Ram* Virginia—*Confederate States Navy*. *Southern Historical Society Papers* XXXII (1896).

Curtis, Richard. *History of the Famous Battle Between the Ironclad Merimac, C.S.N. and the Ironclad Monitor, and the Cumberland and Congress, of the U.S. Navy, March the 8th and 9th, 1862, As Seen by a Man at the Gun*. Norfolk, Virginia: S. B. Turner & Son, 1907.

Eggleston, J. R. *Captain Eggleston's Narrative of the Battle of the Merrimac*. *Southern Historical Society Papers* III, September 1916.

Flanders, Alan B., and Captain Neale Westfall, ed. *Memoirs of E. A. Jack, Steam Engineer, CSS* Virginia. White Stone, Virginia: Brandylane Publishers, 1998.

Foute, R. C. *Echoes from Hampton Roads*. *Southern Historical Society Papers* XIX (July 1891).

Jones, Catesby ap R. *Services of the Virginia (Merrimac)*. *Southern Historical Society Papers* XI (January 1883).

Jones, John B. *A Rebel War Clerk's Diary*, edited by Earl Schenck Miers. New York: A. S. Barnes & Company, Inc., 1961.

Newton, Virginius. *The* Merrimac *or* Virginia. *Southern Historical Society Papers* XX (January 1892).

Norris, William. *The Story of the Confederate States Ship "Virginia" (Once Merrimac)*. *Southern Historical Society Papers*, n.s. 4, (October 1917).

Parker, William Harwar, *Recollections of a Naval Officer, 1841–1865*, New York: Scribner, 1883.

Phillips, Dinwiddie B. *Notes on the Monitor—Merrimac Fight*. *Battles and Leaders of the Civil War*, vol. 1, R. U. Johnson and C. C. Clough Buel, ed. New York: The Century Company, 1887.

Porter, John Luke. *The Plan and Construction of the Merrimac*. *Battles and Leaders of the Civil War*, vol. 1, R. U. Johnson and C. C. Clough Buel, ed. New York: The Century Company, 1887.

Wise, John S. *The End of an Era*. Boston and New York: Houghton Mifflin Company, 1899.

Wood, John Taylor. *The First Fight of Ironclads*. *Battles and Leaders of the Civil War*. volume 1, R. U. Johnson and C. C. Clough Buel, ed., New York: Century Company, 1887.

Woodward, C. Vann, ed. *Mary Chesnut's Civil War.* New York: Book-of-the-Month Club, 1994 reprint of Yale University Press edition, 1981.

Worden, John L., Samuel Dana Greene, and H. Ashton Ramsey. *The Monitor and the Merrimack: Both Sides of the Story.* New York: Harper and Brothers, 1912.

Secondary Sources

Besse, Sumner B. *C.S. Ironclad Virginia and U.S. Ironclad Monitor with Data and References for Scale Models.* Newport News, Virginia: The Mariners' Museum, 1996.

Brooke, George M. *John M. Brooke, Naval Scientist and Educator.* Charlottesville, Virginia: University Press of Virginia, 1980.

Campbell, R. Thomas, and Alan B. Flanders. *Confederate Phoenix, The C.S.S. Virginia.* Shippensburg, PA: Burd Street Press, 2001.

Dew, Charles B. *Ironmaker to the Confederacy, Joseph R. Anderson and the Tredegar Iron Works.* New Haven and London: Yale University Press, 1966.

Durkin, Joseph T. *Stephen R. Mallory: Confederate Navy Chief.* Chapel Hill: University of North Carolina Press, 1954.

Flanders, Alan B. *The Merrimack.* No publisher listed, 1982.

Luraghi, Raimondo, *A History of the Confederate Navy.* Translated by Paolo E. Coletta. London: Chatham Publishing, 1996.

Pollard Edward A. *Life of Jefferson Davis, with a Secret History of the Southern Confederacy,* Philadelphia, PA: National Publishing Co., 1869.

Still, William N. *Confederate Shipbuilding.* Athens: University of Georgia Press, 1969.

———. *Iron Afloat: The Story of the Confederage Ironclads.* Columbia: University of South Caroplina Press, 1985.

———, ed. *The Confederate Navy, the Shops, Men and Organization, 1861–65.* London: Conway Maritime Press, 1997.

Sharf, J. Thomas. *History of the Confederate States Navy.* New York: Fairfax Press, 1977 reprint of the 1887 edition.

Symonds, Craig L. *Confederate Admiral, The Life and Wars of Franklin Buchanan.* Annapolis, Maryland: Naval Institute Press, 1999.

Wells, Tom H. *The Confederate Navy: A Study in Organization.* Alabama: The University of Alabama Press, 1971.

Articles

San Francisco Chronicle. "Story of August Peterson, Captain of No. 9 Gun on the Merrimac." March 15, 1903.

Still, William N. *The Confederate States Navy. Maritime Life and Traditions* 18 (Spring 2003).

Monitor

Manuscript Collections

Greene, Samuel Dana, Letter to Mother and Father. March 14, 1862, Transcript in Mariner's Museum Collection, Newport News, Virginia.

Harbeck Collection. Henry E. Huntington Library, San Marino, California.

National Archives, RG 45, File 1, Box 49. Washington, D.C.

Isaac Newton Papers. Mariners' Museum, Newport News, Virginia.

Welles Collection. Henry E. Huntington Library, San Marino, California.

Worden Papers. M. S. Worden, Box File, Mariner's Museum, Newport News, Virginia.

Published Primary Sources

Basler, Roy P., ed. *The Collected Works of Abraham Lincoln*. New Brunswick, New Jersey: Rutgers University Press, 1953.

Bushnell, Cornelius, et al. *Negotiations for Building the* Monitor. *Battles and Leaders of the Civil War*. Volume 1, R. U. Johnson and C. C. Clough Buel, editors. New York: Century Company, 1887.

———. *Paper read to the Army and Navy Club of Connecticut, June 22, 1894*, reprinted in *Program of Exercises at the Unveiling of the Memorial to Cornelius Scranton Bushnell, New Haven*, May 30, 1906. No publisher given.

Butts, Francis B. *The Loss of the Monitor. Battles and Leaders of the Civil War*, vol. 1, R. U. Johnson and C. C. Clough Buel, ed. New York: The Century Company, 1887.

Daly, Robert W. ed. *Aboard the U.S.S.* Monitor; *1862 The Letters of Acting Paymaster William Frederick Keeler, U.S. Navy, To his Wife, Anna*. Annapolis, Maryland: United States Naval Institute, 1964.

Durbrow, Julia Stimers, ed. *The Monitor and Alban C. Stimers*, Orlando, FL: J. S. Durbrow, 1936.

Ericsson, John. *The Building of the* Monitor. *Battles and Leaders of the Civil War*.

Fox, Gustavus Vasa. *The Confidential Correspondence of Gustavas Vasa Fox*, Edited by Robert Means Thompson and Richard Wainwright. Freeport, New York: Books for Libraries Press, 1972 (reprint of 1920 edition).

Greene, Samuel Dana. *In the "Monitor" Turret. Battle and Leaders of the Civil War*, vol. 1, R. U. Johnson and C. C. Clough Buel, ed. New York: The Century Company, 1887.

Marvel, William, *ed. The Monitor Chronicles, One Sailor's Account*. New York: Simon & Schuster, 2000.

Newton, Isaac. *Shot-Proof Vessels—Ericsson's Battery. Journal of the Franklin Institute of the State of Pennsylvania for the Promotion of the Mechanical*

Arts 73, no. 434 (February 1862). Philadelphia: Published by the Franklin Institute at their Hall.

Ould, Robert. "The Exchange of Prisoners." In *Annals of the War, Written by Leading Participants*. Philadelphia: The Times Publishing Company, 1879.

Porter, David Dixon. *Incidents and Anecdotes of the Civil War*. New York: D. Appleton and Company, 1885.

Stodder, Louis. "Ten Months in the Monitor." *Everybody's Magazine*, December 1900.

Welles, Gideon. *The Diary of Gideon Welles*. Edited by Howard K. Beal. New York: W.W. Norton & Company, 1960.

————. "The First Iron-clad Monitor" in *Annals of the War, Written by Leading Participants*. Philadelphia: The Times Publishing Company, 1879.

Wells, William S., ed. *The Story of the Monitor—The First Naval Conflict Between Ironclad Vessels in Hampton Roads, March 9th, 1862*. New Haven, Connecticut: Issued by the Cornelius S. Bushnell National Memorial Association, 1899.

Wilkes, Charles. *Autobiography of Rear Admiral Charles Wilkes, U.S. Navy, 1798–1877*. Edited by William James Morgan et al. Washington, D.C.: Naval History Division, Department of the Navy, 1978.

Worden, John L., Samuel Dana Greene, and H. Ashton Ramsey, *The Monitor and the Merrimack: Both Sides of the Story*. New York: Harper and Brothers, 1912.

Secondary Sources

Bennett, Frank M. *The Steam Navy of the United States*. Pittsburgh: Warren & Company, 1896.

Besse, Sumner B. *C.S. Ironclad* Virginia *and U.S. Ironclad* Monitor *with Data and References for Scale Models*. Newport News, Virginia: The Mariners' Museum, 1996.

Boynton, Charles B. *The History of the Navy during the Rebellion*, 2 vols. New York: D. Appleton and Company, 1867.

Church, William Conant, *The Life of John Ericsson*. New York: Charles Scribner's Sons, 1911.

Konstam, Angus, Tony Bryan, illustrator. *Union Monitor, 1861—65*. Oxford, England: Osprey Publishing, 2002.

McCordock, Robert Stanley. *The Yankee Cheese Box*. Philadelphia: Dorrance and Company, 1938.

Niven, John. *Gideon Welles, Lincoln's Secretary of the Navy*. New York: Oxford Univesity Press, 1973.

Sandburg, Carl. *Abraham Lincoln, The War Years*. New York: Harcourt Brace & Co., 1939.

Soley, James Russell. *The Union and Confederate Navies. Battles and Leaders of the Civil War*, Volume 1. R. U. Johnson and C. C. Clough Buel, editors, New York: Century Company, 1887.

Weise, Arthur James. *Troy's One Hundred Years, 1789–1889*. Troy, New York: William H. Young, 1891.

West, Richard S. *Gideon Welles, Lincoln's Navy Department*. New York: The Bobbs-Merrill Company, 1943.

White, Ruth. *The Yankee from Sweden, the Dream and the Reality on the Days of John Ericsson*. New York: Henry Holt and Company, 1960.

Articles

"Iron-clad Vessels." *Harpers New Monthly Magazine* XXV, no. CXLVIII, (September 1862).

McCord, Charles. "Ericsson and his Monitors." *The North American Review* 149, issue 395 (October 1889).

Milligan, John D., ed. "An Engineer Aboard the Monitor by Alban Stimers." *Civil War Times Illustrated* IX. (April 1970).

"The Ericsson Battery." *Scientific American*, n.s., 5, issue 21 (November 23, 1861).

Wegner, Dana. "Ericsson's High Priest." *Civil War Times Illustrated* XIII (February 1975).

Government Documents

37th Congress, 2nd session. Senate Report No. 37, *Surrender and Destruction of Navy Yards, Etc.*

40th Congress, 2nd session. Executive Document No. 86, *Letter of the secretary of the Navy Communicating, In compliance with a resolution of the Senate of the 24th instant, information in relation to the construction of the ironclad* Monitor.

Senate Journal, 37th Congress, 1st session, 1861, chapter XXXVIII.

Still, William N., *Monitor Builders: A Historical Study of the Principal Firms and Individuals Involved in the Construction of the U.S.S. Monitor*, Washington, D.C.: Division of History, National Park Service, Department of the Interior, 1988.

General

Manuscripts

Monaghan Papers. Mariners' Museum, Newport News, Virginia.

Published Primary Sources

Coffin, Roland F. *"The First Fight Between Ironclads." Outing* X (August 1887).

Colston, R. E. *"Watching the Merrimac." Battles and Leaders of the Civil War,* vol. 1, R. U. Johnson and C. C. Clough Buel, ed. New York: The Century Company, 1887.

Volume 1, R. U. Johnson and C. C. Clough Buel, editors, New York: Century Company, 1887.

Curtis, Frederick H. *"Congress and the Merrimac." New England Magazine* XIX (February 1899).

Nevins, Allan, and Milton Halsey Thomas, eds. *The Diary of George Templeton Strong, The Civil War Years, 1860–1865.* New York: The Macmillan Company, 1952.

Reaney, Henry, Acting Master, U.S.N. *"How the Gunboat 'Zouave' Aided the 'Congress.'" Battles and Leaders of the Civil War.* Volume 1, R. U. Johnson and C. C. Clough Buel, editors, New York: Century Company, 1887.

Selfridge, Thomas O. *"The Merrimack and the Cumberland." The Cosmopolitan* XV (1893).

———. *Memoirs of Thomas O. Selfridge, Rear Admiral, U.S.N.* New York: G. P. Putnam's Sons, 1924.

Shippen, Edward. *Thirty Years at Sea.* Philadelphia: J. B. Lippincott & Co., 1879.

———. *"A Reminiscence of the First Ironclad Fight." Lippincotts Magazine* XXI (1878).

Swinton, William. *"The Monitor and the Merrimac."* In *The Twelve Decisive Battles of the War.* New York: Dick and Fitzgerald, 1867.

Secondary Sources

American National Biography, Volume 5. New York: Oxford University Press, 1999.

Baxter, James Phinney. *The Introduction of the Ironclad Warship.* Cambridge, MA: Harvard University Press, 1933.

Davis, William C. *Duel Between the First Ironclads.* Baton Rouge, Louisiana State University Press, 1981 (reprint of 1975 edition).

Denney, Robert E. *The Civil War Years: A Day by Day Chronicle*. New York: Gramercy Books, 1998.

Gosnell, Allen H., *Guns on the Western Waters: The Story of the River Gunboats in the Civil War,* Baton Rouge: Louisiana State University Press, 1949.

McPherson, James. *Battle Cry of Freedom*. New York: Oxford University Press, 1988.

Musicant, Ivan. *Divided Waters*. New York: HarperCollins Publishers, 1995.

Paine, Lincoln P. *Ships of the World*. Boston: Houghton Mifflin Company, 1997.

————. *Warships of the World to 1900*. Boston: Houghton Mifflin Company, 2000.

Smith, Gene A. *Iron and Heavy Guns: Duel Between the* Monitor *and* Merrimac, Abilene, Texas: McWhiney Foundation Press, 1998.

Government Documents

Navy Department. *Civil War Chronology, 1861–1865*. Washington, D.C.: Naval History Division, 1971.

Official Records of the Union and Confederate Navies in the War of the Rebellion, Series I, 27 vols.; Series II, 3 vols. Washington, D.C.: 1880–1902; reprint: 1999, Carmel, Indiana, Guild Press of Indiana, CD-ROM.

Index

BOOKS BY JAMES L. NELSON

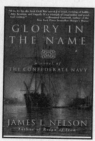

GLORY IN THE NAME
A Novel of the Confederate Navy
ISBN 0-06-095905-3 (paperback)
"By far, the best Civil War novel I've read;
reeking of battle, duty, heroism, and tragedy.
It's a triumph of imagination and good, taut
writing." —Bernard Cornwell

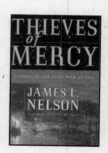

THIEVES OF MERCY
A Novel of the Civil War at Sea
ISBN 0-06-019970-9 (hardcover—April 2005)
Filled with wild characters, heart-pounding
action, and set against the bold backdrop of
the Civil War, *Thieves of Mercy* is a worthy
sequel to *Glory in the Name.*

THE GUARDSHIP
Book One of the Brethren of the Coast
ISBN 0-380-80452-2 (paperback)
A threat from Marlowe's illicit past as a
pirate looms on the horizon, and Marlowe
must choose between losing all or facing the
one man he fears.

THE BLACKBIRDER
Book Two of the Brethren of the Coast
ISBN 0-06-000779-6 (paperback)
"[Nelson's] descriptions have the ring of
truth and are conveyed with a sharpness
and clarity that even the landbound can
appreciate." —*Chicago Tribune*

THE PIRATE ROUND
Book Three of the Brethren of the Coast
ISBN 0-06-053926-7 (paperback)
"A rousing swashbuckler filled with treasure,
sea battles, feuds, revenge, romance, and
deadly conspiracies. . . . A full broadside of
reading entertainment." —*Publishers Weekly*

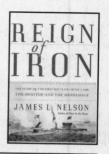

REIGN OF IRON
*The Story of the First Battling Ironclads,
the Monitor and the Merrimack*
ISBN 0-06-052403-0 (hardcover)
Nelson's rousing first non-fiction book,
on one of the great naval battles that
was a turning point in U.S. history.